The Mystery of Everything

From Experimental Sciences and Philosophy to Revelation and Faith

OSVALDO DELLA GIUSTINA

Copyright © 2022 OSVALDO DELLA GIUSTINA
Copyright © 2022 Generis Publishing

All rights reserved. This book or any portion thereof may not be reproduced or used in any manner whatsoever without the written permission of the publisher except for the use of brief quotations in a book review.

Title: The Mystery of Everything

From Experimental Sciences and Philosophy to Revelation and Faith

ISBN: 979-8-88676-094-1

Author: OSVALDO DELLA GIUSTINA

Cover image: www.pixabay.com

Publisher: Generis Publishing
Online orders: www.generis-publishing.com
Contact email: info@generis-publishing.com

"Knowledge of the experimental sciences gives Faith its dimension related to quantitative things. Philosophy seeks its nature, or essence. But Faith gives Science and Philosophy its transcendent dimension. In this way they harmonize and complete each other."

OSVALDO DELLA GIUSTINA

"Professor Osvaldo- When I read his book "O Mistério de Tudo" I was convinced enough to understand that it is a monument to his own intellectual and personal maturity... The ultimate connoisseur does not exist, and a final theory or everything is not viable. Mathematics may have formal universal validity, but no mathematician is universally valid. Thanks for everything".

Pedro Demo, teacher, writer, doctor in Education

"The book "The Mystery of Everything" for those who lived with the author for long years, rather than, as he claims, constituting a conclusion of his long bibliography written in books, articles, lectures and, especially, on the social networks, constitutes a synthesis of his life, dedicated to the construction of a better world, which he calls Participatory and Solidarity, or, disciple I confess, that it is, by Teilhard de Chardin, from Amorized Civilization. In response to his invitation, it is worth taking this path with him."

João Jerônimo Medeiros, Professor Emeritus at Uni

Osvaldo Della Giustina, professor, philosopher, writer and academic, From Santa Catarina, with the book "O MISTÉRIO DE TUDO", crowns his invaluable work in the field of higher education, in public life and in literature, with intelligent reflections and high philosophical questions, theological and cultural, on the meaning of man, life and faith. This one extraordinary work, for the very high level and depth of his analyses, puts the work on a higher level of human interest, for the Brazil and for the world.

**Moacir Pereira, Journalist, Writer,
President of Academia Catarinense of Letters**

THE MYSTERY OF EVERYTHING
OSVALDO DELLA GIUSTINA

THE REASONS OF MY FAITH

THE MYSTERY OF EVERYTHING

Virtual platforms: www.participacaoesolidariedade.com.br

Osvaldo Della Giustina
della_giustina@terra.com.br

Valério Azevedo, assistant
vanaweb@gmail.com

Milenne Kelly, audiovisual producer
milennek@gmail.com

Translation into English by Emerson Gontijo Penha, Belo Horizonte, Brazil, 2021
epenha@horizontefilmesbrasil.com.br

People who kindly contributed on the previous reading of the text: Pedro Demo, João Jerônimo Medeiros and Moacir Pereira.
To those and others who somehow collaborated, I send my gratitude.
All responsibility for the concepts, analysis and conclusions is on my exclusive liability.

Brasilia, DF, Brazil - 2020/2021

To Aurora, my wife,
to Jaison, Leslie, Yuri, Glauber, Christian, my kids,
for my sons and daughters-in-law,
and to the new generations

There are a large number of sacred books, consecrated by different peoples and cultures, as well as different versions of the Bible. The biblical texts cited in this book were taken from the Holy Bible, translated from the originals through a version by the monks of Maredsous, Belgium, and rendered into Portuguese by the Catholic Biblical Center, edited in its 52nd Edition by Editora Ave Maria, São Paulo, 1957. However, in this English version, we refer to the New Revised Standard Version, Catholic Edition, National Council of Churches, USA, first published in 1989.

PRESENTATION

When the human species appeared in remote times, the process of evolution took a leap in the way it had been happening. Animal species were born from dust, procreated, and returned to dust, without any noticeable transformation from one generation to another.

According to the author of "The Mystery of Everything", the fact is that, suddenly, a new species in this process, developed a new way of being and living, created a new way of communicating, a new language and, the essential, began to perceive itself as an independent individual, creating consequently codes of conduct, separating good from evil, finally starting to exist in another sphere, or in another world that definitively separated it from other species. This species had been endowed with a new attribute: Consciousness.

For the questions raised since then and brought to reflection until today, millions of years later, and even knowing that *"in everything there is the Mystery,"* the author turns to the search for answers, urging the reader to follow him in this long and complex journey: how or why did this transformation take place? Where did Consciousness come from, where did we come from and where are we going? What is the universe around us, where does it come from? What are we, after all, *creatures in the image and likeness* of those who created us? Has there been a Creator and has he created us to suffer and make others suffer? How can one explain that aiming the good, we often do the bad? Will everything be by chance, or will there be a project that gives meaning to all creation, to the universe and everything it contains, to man and his destiny? If there is evil, or if evil exists and is a constant threat to the entire human species, is there a project of redemption, so that good can prevail over evil? If we don't see it, where is it, or what will this project be?

These are all questions whose answers, or unfathomable mysteries, the reader will be able to discover along with the author who, as he insists on saying, is absolutely aware that the options regarding each answer or each mystery, belong to each one, but he has the certainty, also, that each one who makes this path with him, will leave seeing greater horizons, or with enriched minds and hearts.

However, this process of questions and answers contains within itself a dialectic, in which each answer generates a new question, and it is for this reason that the book seeks to add the conclusions of the experimental sciences, in which it stands as a "contemporary apprentice" of theology, in which he puts himself as an "eternal neophyte", but claiming for his essential condition of "philosopher, whose logic" helped him to respond "to the revelation" and **assume and preserve the faith**.

In this effort, the author imagines creating a harmony in which each of these knowledges contributes in something to the growth of the other: *"Knowledge of the experimental sciences gives Faith its dimension related to quantitative things. Philosophy seeks its nature, or essence. But faith gives Science and Philosophy their transcendent dimension. In this way they harmonize and complete each other"*, he says, and concludes:

"Experimental knowledge, or philosophy, is not faith. Faith is the acceptance of revealed knowledge, or the Mystery. Faith is the option to believe, to accept, even if you don't understand, or exactly because you don't understand."

The author, therefore, does not see opposition between these various forms of knowledge, and assumes them peremptorily as his options and, confessedly, offers them to the reader, leaving each one to exercise their own options, in accordance with their own conditions, their circumstances, their will, or whatever their conscience dictates.

Philosopher, educator, journalist and writer, Osvaldo Della Giustina is someone who has a vast intellectual repertoire, gathered over more than sixty years of studies and reflections, since his philosophy course, started at the Major Seminary in Viamão, in Southern Brazil, and concluded at PUC University in Porto Alegre, whilst he was practicing in journalism. With a diversified career at all levels of professional performance, in the state of Santa Catarina, where he was born, or in Brasília and with stops in several states and many missions carried out in other countries, thus adding multiple experiences, always exposed and debated, hundreds of academic articles published in Brazil and abroad, he is also the author of two dozen books in Portuguese, some of them translated into English, French and Russian.

With all this accumulated wealth, his proposal for a social organization to contribute to the evolution of mankind itself and its form of coexistence in an increasingly humanized dimension, an effect that he characterizes as Participatory and Solidarity towards the construction of an amorous civilization.

In his bibliography he has come a long way to reach this proposal. He began walking on this path in 1982, with the book *THE AGE OF MAN, Fundamentals for a New Social Order*, followed in 2000, at the turn of the century, with *For Humanization of Society, THE THIRD MILLENNIUM REVOLUTION* and in 2004, with *PARTICIPATION AND SOLIDARITY, The Third Millennium Revolution (II)*. Finally, in 2019, he offered an elaborate synthesis of this trilogy with *For a Participative and Solidary Civilization, THE PROPOSAL.*

Now 'Professor Osvaldo', as he is affectionately called by those who have the privilege of enjoying his conviviality, completes his work with this reflection on man or the mankind, aware that social structures, or civilizations, are a product of the human species, and so they are the expression of the human beings that constitute it. So, as he himself confesses, in a way like "crowning his work", he presents his readers with this new challenge, which is THE MYSTERY OF EVERYTHING: *of the Experimental Sciences and from Philosophy to Theology, to Revelation and to Faith*, a title he completes, as in a confession: *The Reasons of My Faith.*

The Mystery of Everything is an instigating combination of the conclusions of Experimental Science, with the logic of Philosophy and the contents of Revelation from both Bible and Theology, which added to the whole process of evolution his connection to Faith.

With this multiple approach, in which the author moves with aplomb, the book makes a significant contribution to the ever-present debate on the relationship between scientific knowledge and faith and guides the reader on an exciting journey through a new way of thinking about the mysteries of life, man and the universe.

At the end of the book, Osvaldo, the humanist, confesses his emotion on walking along two different roads at once, one of the trilogy on the organization of civilization and its synthesis in *PROPOSAL (walking along the path of experimental science, with a little philosophy)*, and another that leads to this "MYSTERY OF EVERYTHING" *(coming through faith, through the paths of*

Theology and Revelation, with a lot of logic in philosophy). These two paths happened to meet at the same point, or towards the same goal: the construction of a fairer and more humane society, which he calls more Participatory and Solidarity, and repeating, towards an **Amorous** Civilization, a term sought in the anthropology of Teilhard de Chardin, whom he considers to be his master.

For the author, this meeting of science and faith in an amorous civilization demonstrates that the advent of a new civilization for the world to come is inevitable. And it will necessarily come different from everything that has been seen or experienced in history and will perforce come after the advances of Science and Technology from the values of the Mass of Consciousness, as he calls them, values that coincide with the Message of Christ, the Redeemer, which placed at its center, or in its essence, the greatest of all commandments: *The Love that sums up all the law and the prophets.*

Finally, reinforcing the Author's warning, THE MYSTERY OF EVERYTHING is not a book of religion, in the traditional sense of the word, but a reflection that led him (and it is now offered to those who follow him) to penetrate the essential meaning of life, of human species and the universe, at this time of evolution when Civilization seeks new steps to build the next levels in the history of the mankind and civilization.

Valerio Azevedo
Journalist and Writer
Brasília, Brazil

THE MYSTERY OF EVERYTHING
(FROM EXPERIMENTAL SCIENCES AND PHILOSOPHY TO REVELATION AND FAITH)

INTRODUCTION

The reflections I offer below support my faith. I offer them with a brief reference to what the experimental sciences say, with the application of the logic of philosophy and with the view of theology, forms of knowledge that I consider coherent with each other, the experimental sciences, with the logic of philosophy and with the theology that interprets us or leads us to Revelation and the Revelation that grounds our faith.

Revelation, therefore, must be considered an instrument or a method of knowledge, which is not to be confused with the knowledge of experimental science, or with the logical knowledge of philosophy, or even with the interpretations of theology. In fact, beyond theology, when we are confronted with the Mystery, or the essences, only in Revelation will we find answers and it is in this dimension that Revelation imposes itself as a foundation of faith.

I hope that such concepts and their ways, well understood, can contribute in some way to those who, together with me, reflect on their own questions. This posture, the inquiring, more than any other attribute characterizes human beings.

To those who inquire, I invite them to follow these reflections, because I consider the searches as important as the encounters or discoveries, especially when the searches concern the essential of our existence, where only indifference would be a less human position, which would be regrettable.

My goal, however, is not to induce anyone to adhere to these concepts and to the options they offer, but it is just to clarify the reasons for my Faith, because the richness of the Civilization that we seek is diversity; and the understanding of the reasons for diversity constitutes the presupposition of harmonious coexistence, and harmonious coexistence in diversity, in turn, is the very vestibule of Love, or of the **Amorous Civilization**, that the human species, or the divine Project of Creation, seeks to build.

Thus, by bringing together the various levels of knowledge, the desire to contribute to the speculation, curiosity, and anxieties of many and to the indifferences of the few, trying to be faithful to the essence of knowledge and faith, although conscious, or precisely because conscious, knowing that in everything remains the Mystery. Or, more properly, knowing that faith exists, it makes sense, and it becomes necessary, because the Mystery exists.

The knowledge of experimental Science, also of Philosophy, is not faith. Faith is the acceptance of the Revealed Knowledge, of the Mystery. Faith is the option to believe, to accept, even if you don't understand, or exactly because you don't understand.

For this reason, **faith is not irrational, nor is knowledge exhausted in experience or logic. The knowledge of experimental Science gives faith its dimension related to quantifiable things. Philosophy seeks its nature or essence, but faith illuminates and gives Science and Philosophy their transcendent dimension. In this way they complete each other.**

Far be it from me, however, one way or the other, to want to explain everything. I only want to invite to a reflection I understand as explainable, open to accepting that in everything there is the inexplicable and that the Mystery is inherent to the limitations of our nature.

My invitation is addressed to everyone, but especially to those who seek to assume this full human dimension, beyond things, to travel the path of elevated knowledge, this path, to its transcendent dimension, beyond things

Throughout these reflections, it was often evident to me that imagining that we know everything, or that any level of science explains everything, and that everything can be unraveled or explained, mainly reveals lack of perception, or lack of faith (of the Grace) in relation to the essentials of things, and how they are in their nature. This would be a mistake like imagining that, by looking at the surface of the waves from the beach, one would know the entire dimension of the ocean - which, of course, would only evidence a hidden satisfaction or pride, blind to the essential, as that the dimension of the ocean can only be understood when, beyond the waves and the surface, one is able to penetrate its depth and extension. While these misinterpretations can be painful to admit, honestly, it must be agreed that they exist, consciously or unconsciously, in all human beings. Unknowingly in some, explicitly in many. Human beings... What are human beings in the face of the complexity of the species, the passage of time, space and eternity or the immensity of the universe, or what is its opposite, nothingness? Or could there be something beyond time and space?

As we consider human beings, the human species, the only conscious beings in the universe (which they may be, although we have no clear proof), this solitary existence would be a first Mystery. It is certain that, according to the Book of Genesis, the first book of the Bible widely spread and accepted by most cultures, God gave man the earth and everything in it so that he could occupy the land, take care of and complete it, giving continuity to the work of creation.

In this perception, our existence as the only conscious species on Planet Earth (or would we be unique in the universe?) would mean giving human beings an enormous dimension and an immense responsibility. I say more.

It would surely mean that we are only in the prehistory of the human species and that would force us to ask what we will be in the future, or what the human species will be in the future. This hypothesis, that of being the only conscious species in the universe and, therefore, of being only in our prehistory, would also give a new almost transcendental meaning, to the conquest of space, beyond the Planet's borders. And I say even more.

As we head into space, outside our Planet, or our system, even though, at this moment, there is not even a practical sense to the question, I insist – what form will we have, or how will we occupy this new universe? What will we take for him, life or destruction, love or hate, war or peace, messengers that we are, *made in the image and likeness of the Creator* (Genesis 1:26) who wanted to associate us with his universal project of creation? What kind of seed will we be? From this hypothesis and so many others, a series of questions arise.

- Would this species, the human species, **be large enough** to assume responsibility for the Universe? Or would God have imagined, *by creating us in his image and likeness*, to share this responsibility with mankind and, therefore, to give us the conditions and the corresponding means? What would these conditions be? What means would we have? Or would God leave everything for man to discover and develop what is necessary for his improvement and the improvement of the Planet? Or of the Universe? Or would he have left, the Creator, the necessary instruments, or some project to be fulfilled with this dimension? Or would it be enough to give him Consciousness?

- **The Universe**... where did the universe come from? Where will it go? When did it start and what are its limits? Or is it that it has never started, has no limits, and will exist by itself, with no end or beginning?

- **What does the human species mean** in the universe and for the universe? Where did this species come from and where is it going? How do experimental science, philosophy, or theology, revelation and faith stand before the human dimension? Can such forms of knowledge coexist harmoniously? Is it possible to get an answer from them?

- **If admitted that the universe had no** beginning and will not have an end or limits, this would not lead us to admit that it - the universe - does not have the essential characteristics of all known matter: form, time and space, the beginning and the Finally, as part of everything that can be seen, felt, touched (or at least interpreted even if only by mathematical equations that define it). In this case, if the Universe is beyond these characteristics, would not it have the same nature attributed to the spirit? Isn't the universe its own god?

- **I do not see how this contradiction** of a material being that should deny its own nature ceasing to be what it is in order to explain itself can be accepted. This impossibility forces us to admit the existence of an absolute Being in another order, the order outside the universe of time and space, which means a Being by nature essentially independent of space, time and, therefore, absolute in this order, the order of the Spirit, can this being explain the origin of the universe?

- **This admitted, how to accept the conception of the Bible**, that this Being of another order to whom it is possible, for being of another order, to attribute the creation of the universe, has conceived a species *"in its image and likeness"*, but that, despite of this similarity, suffers and is capable of causing suffering, does the good, but does the evil? If evil somehow inserts itself in the nature of this species, made in the image and likeness of its Creator, would it be justified for the Creator to let this species get lost forever, or logic would force us to expect, or admit, the existence of a project of Redemption, capable of saving the creation of such a tragic fate?

- **If logic leads us to admit the existence of this Redeemer Project**, how would this project have to be? how would it take place? what and how would the participation of the human species be in it? after all, what is the meaning of the human species, where is this species heading, since its appearance in the Project of Creation until the future, or until the end of time if this is its destiny and, in this perspective, what is the meaning, or the final destiny for the universe itself? or will there be neither destiny nor finality for the species or the universe, and will everything remain so indefinitely, without explanation? Just unanswered questions?

- **Or is it worth the search for answers?** In this case, how far will the answers be possible, and where will the logic of acceptance of the Mystery begin and, consequently, **in the search for answers, must the Mystery always impose itself?**

- **Finally, where does the search** for experimental science and philosophy, theology or revelation and faith begin, how does it develop and end? Where is each one's space established and where is the space for collaboration or complementation among themselves, if there is such space in this common search? Or will this search never be common?

As you can see, there are many questions. Even more considering that in addition to these, there are many others that, as we will see, will emerge as we walk in this reflection, or in this search...

In summary, the objective of this book, together with those who wish to seek answers to the world of essences and transcendences, that is, what goes beyond what appears, surfaces and their hidden Mysteries, as well as those who accept to penetrate the limits that Consciousness allows us **is the encounter with the Mysteries of everything** and, going beyond the experimental sciences and philosophy, the search for theological answers or the encounter with *revealed knowledge*, source of faith, its acceptance, or its rejection.

The following text considers, therefore, that all these forms of knowledge need to be considered, not necessarily to get answers to everything, but to give the real dimension of what can be understood and of the Mysteries that are around us and of which we are part, or in which we are immersed. In this way, the idea is that we can understand ourselves a little better about how we really are, what the universe is like, where we live and what meaning it has, or how far theology, revelation and faith can take us. Or what horizons can they open to us, beyond those that limit our purely quantitative, material vision?

I say, however, that it would be a mistake to see in these reflections a book on theology, or on religion. I want this book to be just an introduction to knowledge of a transcendental character capable of opening horizons... opening horizons.... About faith, I refer only to my individual option, possessed and expressed as a confession, or as a showcase offered to those who want to look at it, know it and share it or not. Agreeing, accepting or not, is an absolutely individual option, because, although there is only one truth, to reach the truth there may be several paths. However, it would be better if the best paths were chosen. In this **book I offer the choice of my "way, truth and life", which means, I offer the reasons for my faith.**

On faith itself, I conclude once again with the warning that it is not a consequence of the knowledge of experimental science or even of philosophical logic. To accept what we know through science, we don't need faith, science is enough. We need faith to accept what we do not know through science, where science does not arrive, in other words, before the Mystery.

Where Science does not reach, we are offered Revelation, of which theology is the interpreter, another kind of knowledge that comes from the One who, being beyond all that can be experienced, opens us the horizons of faith. For those who accept it, faith will help them to know, or illuminate the paths of existing and being, and it will certainly open up horizons, broad and new horizons...

Upon discovering, or better yet, upon coming to understand, even if minimally, science, the various levels of knowledge, including revelation and faith, it will be seen that each of these spheres of Consciousness, in the limitation of each one, can strengthen each other. We will then have consciences, or human beings to the fullest extent.

I want to say, however, that this enrichment only happens if, as a presupposition, we are open, free of prejudices towards one or the other, science or faith. And so much more will we receive from them, in knowledge or in vision of horizons, if we seek them, both science and faith.

Personally, I also want to say that, in relation to the experimental sciences, I am only an eternal apprentice; in theology I am no more than a permanent neophyte. I claim, however, the essential condition of a philosopher, which helped me to respond to the revelation and to assume and persevere the faith, exposing it in this window.

Finally, I note that, for the success of this search, there is sometimes a need for sufficient humility. The French writer Leon Bloy (1846-1917), one of the great converts to the Catholic faith at the beginning of the last century, used to say: *"My faith is that of a Breton peasant. But if my faith were absolute and full, it would be like that of a Breton peasant woman"*.

Perhaps we should only arrive at the faith of a Breton peasant, without going beyond. But that choice who should make it is not part of this book. The choice is up to each one of you, your search, your acceptance, your rejection, or how far to go.

In any case, I am sure that this book, for those who reflect it free from prejudice, will open paths for a deepening of the meaning of existing, its essential contents and the reasons that justify walking the path and adhering to faith, or not, by choice of the Consciousness of each one, surely enriched by having made the path.

Eventually I hope you can help in this option, or in offering some answer to doubts, or anxieties, if any, the perspective of the wise words of St. John XXIII, the good Pope: *bisogna vedere tutto, fare quel che se puó fare e l'altre cose, lasciare a Dio.*

Osvaldo Della Giustina, may 2021.

PART 1 – THE MYSTERY OF THE UNIVERSE

1. Of the infinity or eternity of Universe

1.1. What Experimental Sciences say

The ancients imagined the universe in the most different ways, primitive as their knowledge was, but which, by then, must have answered their questions, the same ones we still ask ourselves a few millennia later, despite everything that has been seen and discovered.

For the ancient Hindus, for example, the earth was a half-sphere carried by three elephants carried by a turtle, a realization which the Vedas completed by adding that it was carried by an infinite number of elephants. Evidently such perceptions were no more than mere belief based on the sacredness attributed to these animals.

The Egyptians, on the other hand, proposed a more rational solution, believing that the earth came from a huge initial egg that exploded. This cosmogony claimed that other explosions followed, and their splinters formed the universe, the stars being the souls of the dead. Despite the limitations of its scientific foundations, it is impossible not to relate this idea to the modern theory of the Big Bang, the primordial explosion that is supposed to have given rise to the universe.

On the other hand, the Chinese, of the oldest peoples who have bequeathed their philosophical interpretations since millennia before Christ, used the stars for a series of practical knowledge, such as establishing calendars and determining propitious times for sowing or harvesting.

The Greeks, bent on seeking wisdom, looked to philosophy for answers. Plato proposed that the universe was generated by thought giving life to the image, a theory that Aristotle, preferring to affirm the real, completed by proposing that in the limits of the universe there should be some different kind of matter, since nothing, not being able to exist, required to admit the existence exterior of another kind of universe. It is also impossible not to bring this thought closer to the modern theory of black holes, anti-matter, parallel universes or the dimension of spirit, in addition to space and time.

From the 15th and 16th centuries, at the end of the Middle Ages, the Renaissance promoted, in Western Europe, an intense resumption of the philosophical speculations that emerged in Classical Antiquity. The spread of this line of thought progressively softened the influence of religious dogmas and mysticism on culture and society, promoting a growing appreciation of rationality, nature and science.

Throughout this process – in this region of the world and at this time – the human being came to occupy the center of Creation, and in this environment of many doubts and questions, what we call Modern Science emerged, a way to explain natural phenomena from observation and experimentation with the use of instruments such as the telescope, the microscope, and others within which, in current science, the computer and its countless derivatives.

The birth of this new way of interpreting the universe can be considered, in the history of mankind, a civilizing revolution, as in other times was the discovery of agriculture, since in Antiquity and in the medieval period the investigations of celestial phenomena and living organisms, among other things, they did not use the technique and did not perceive the universe as something composed of the same uniform matter, susceptible to entropy and finitude, theories that would only appear centuries later.

Anyway, until the emergence of Modern Science, or the strengthening of experimental sciences, in general, the Earth was always considered to be flat and the center of the Universe. For some, philosophy was the science to which all knowledge was subordinated, especially in Greece, where the tendency was to answer all human inquiries through philosophy.

Meanwhile, in the centers of Muslim culture, in North Africa and Spain, autonomous sciences were already admitted, and from them there were explanations, or alternative proposals, such as, for example, about the sphericity of the earth, the nature of the stars, or the constellations, the existence of planets and other speculations.

This knowledge established the emergence of a new science, astronomy, which, using mathematics already used in ancient times, allowed the measurement of the size of the earth, or the distances between it and other stars, the moon or the sun, and also, their movements. Thus, began to walk in the direction of modern science.

At the same time came the great navigations, which proved, even for the most disbelieving, the circumference of the earth, its movements, and its position in the universe. But the great theories that intend to explain the universe only happened from the 20th and 21st centuries, in our time, therefore, in the perspective of a new science, different from everything that had been seen, or thought, in the past.

In the 1920s a priest and astronomer, Lemaitre, proposed the interpretation of the origin of the universe through an explosion of an absolutely unthinkable dimension, dubbed **the Big Bang**, produced from a nucleus with a weight, or a mass, of equal dimensions. So equally unthinkable, but capable of causing such an explosion.

Driven by this explosion, the universe would continue to expand, generating new stars, new galaxies, new black holes...

For more than a decade Lemaitre's Big Bang theory was not accepted by scientists, including Albert Einstein, who only admitted it after several other futile attempts at interpretation, inserting it into his vision of the universe. This, however, would only happen in the following decade, of 1930, given the evidence of the expansion of the celestial bodies.

Another physicist, also an astronomer, English, Steve Hawkins, in addition to deepening research on gravity and other phenomena of the universe, popularized in his writings the Big Bang theory, which is now accepted by scientists and others who are interested in the subject, as the most likely explanation.
But no physicist, no astronomer, no scientist has managed to come up with an answer definitively accepted by science, despite multiple attempts, such as the parallel universes, the cycles of expansion and retraction of matter, to a definitively accepted answer, he said, about the origin of this initial nucleus, or the fate of the expanding universe, which remains unexplained.

Meanwhile, the development of experimental science led to the discovery of the atom and its components until reaching the most intimate part of matter, its smallest particle; the development of telescopes, penetrating the world of stars to the observable or calculable limits of the material universe; finally, the penetration of things as they can be captured, whether through telescopes, mathematics, applied algorithms, or imagination. Even so, the experimental sciences have never managed to exceed the limits of time and space, which is their limit, that limits even when in the second degree, I refer to the second-degree abstraction, the level of mathematics. But human consciousness, would it have its limits, would it be satisfied, there?

Today, astrophysics pollutes space with satellites, ships, and other equipment, and the human species visits the moon, research instruments land on Mars, circle Saturn, or penetrate space at the extremes of the galaxy, searching, in light years, for those spaces... but the limits remain...

The essential questions remain, then:
- **where** did the initial story come from?
- **where** did the core of matter and its energy, origin of the initial explosion come from?
- **where** did the parallel worlds come from?
- **in short**, what are we, where did we come from, where are we going, us and the Universe?

Despite all the advances, or perhaps each advance, it seems that the questions multiply and for each answer, a new door opens to a new Mystery.

1.2. Will there be an answer to the new questions?

Up to this point, we have passed our eyes on where the answers from history, astronomy and science arrived, at this point in which it opens the paths of space for the human species. However, experimental sciences, these, or others, are unable to seek answers beyond the universe of mass, shape, weight, or else, time and space, or beyond quantity. It is normal that it be that way because this limitation is of its nature, as it is of human nature.

In fact, to reach answers, or at least orient meanings to the new questions, or the new Mysteries, to take a step beyond the science of mass, shape, weight, time and space, it is necessary to go beyond the quantitative, therefore, beyond physics, history, chemistry or biology, which is to say, going to the world of metaphysics, or philosophy, where only Consciousness, reasoning or pure logic (not experientable) can indicate paths and eventually there can be answers. Further, only the Mystery or the Revelation, possibly with the help of theology. This is also a dimension of human nature.

At this point, therefore, the experimental sciences, including mathematics, come to an end, and pure thought enters, the logic that is capable of verifying the conformity of one concept with another, or of an assertion of knowledge with another, or with its objective, as method of seeking knowledge. Therefore, at this level, philosophy is considered a science of the third degree of abstraction, in addition to the experimental science of the first degree and beyond mathematics, of the second degree of abstraction.

Revelation, however, is still beyond mathematics and philosophy. It is not an experientable science, but its existence scientifically imposes itself, when we reach beyond what can be explained experimentally, mathematically, and even logically, when we reach the Mystery. It is at this level that it is necessary, as a consequence, to accept the Revelation.

Scientists, especially those linked to physics, quantum theory, parallel universes, black holes, astrophysics, do well, putting themselves in their place, by recognizing that the impossibility of answering what was there before the beginning remains, or beyond the extreme limit, of time and space respectively, that this level is not in their scope of knowledge or their scientific method. This is also true for biologists and their searches for the last secrets of life... for this reason, although it is easier to deny, there is no way to rule out that this knowledge, in addition to the quantitative, the level at which they are located, exists.

Considering, however, that not everyone has this perception, it is important to state that experimental science, or other levels, lacks scientific or methodological competence, to say that there is no knowledge beyond theirs, and that there is nothing to do beyond the experiential limits, not knowing or denying, that there can be a science or a new kind of knowledge, or an object (of science), a Being, beyond their experientable objects or their equations, therefore, knowledge, or a science that goes beyond space and time.

Now, if there is this kind of revealed knowledge, the existence of a Being that reveals it becomes necessary, which, as its object, is beyond space and time, not quantifiable or non-equatable. It is also necessary that there is a method in addition to those used for the knowledge of time and space, the equations of mathematics, or the logic of philosophy. Finally, it must be concluded that this revealed knowledge includes, in a process of self-revelation, the revelation of the very nature of this Being and Its attributes, and so only limited, the revelation, by the limitation of the limited nature of the human species, which receives it.

It does not seem to me, therefore, scientific to deny this kind of knowledge, which is imposed by the limits of our own knowledge and more, the limitations of the human species itself, knowledge, and species which, however, can, yes, be expanded, as in which you accept the revealed knowledge.

Consequently, it must be concluded that this acceptance is the very essential object of faith, which also expands with the understanding of the richness of the multiple ways in which our limitations, including knowledge, can be overcome...

Regardless, therefore, of those who deny or are unaware of the method of knowledge beyond space and time or the quantitative, but along with those who know that there is scientific knowledge beyond quantitative methods, first recognizing the logical method of philosophy and, beyond logic of philosophy, to seek an answer to essential unanswered questions, from those who know that there is, as said, a legitimate knowledge of another level, coming from revelation, from which theology also derives, which studies it, which seeks to interpret it and apply it. Its legitimacy lies in the difference from method, or from experimental knowledge, and the acceptance of difference is an essential part of the method of science, as of knowledge.

Considering this, we reach the limit of time and space, and it is worth asking: what can exist beyond? Let's see if the logic substantiates this existence that, if proven, will take us forward, to the legitimacy of the Revelation.

a) **The first logic is that existence is possible beyond space and time.**

If we conceive in knowledge a world where space and time do not exist, we are in another order of being, obviously a being that exists without time and without space. To exist without time and without space means to be eternal and infinite, because there is no time, there is no space, which means being of a different nature than the nature of experienceable beings. Whatever is experienceable, therefore, will not reach this other order.

Thus, the existence of the world beyond time and space - the world beyond physics and other sciences of that order, the world of metaphysics - or philosophy, where there is no way to apply experimental science, including mathematics, 2nd degree knowledge (of abstraction). The existence of this timeless and spaceless world therefore presupposes the existence of a method that is of the same nature as its object, that is, logic, not experimental logic or even mathematical logic and its equations, but pure logic, which harmonizes its method with its object.

It is this method and the nature of its object that is independent of time and space that gives consistency and defines a science also of another nature, on another level, philosophy, or metaphysics.

It is this pure logic, science of another nature, that indicates the existence of a Being of another order, of another nature, a Being that does not depend on the matter of which the universe was made. Therefore, not depending on matter to exist and being able, as a consequence, to exist before the existence of matter, this Being, according to its nature, can constitute the cause capable of explaining the origin of a world that experimental science, limited to time and space, does not reach. Coincidentally, it can be seen below how this reasoning coincides with the same logic as Saint Thomas Aquinas, when exposing his 5 ways through which knowledge of the existence of God is reached.

This Being of another order constitutes, therefore, the very nature of the spirit, being without space and without time (without weight, without form, without mass, etc.), a being, therefore, immaterial, absolutely different from the nature of the material universe, a being eternal because it has no time and a being that is everywhere, or in our concept of part, nowhere, because it has no space.

b) The second logic: the capacity of the immaterial being to communicate with the material being.

Given that this Being from another order of material being exists, a second logical deduction follows, which answers the following question.

Can this being of another order, of a purely immaterial order, relate to this other universe of experimental things, conditioned by time and space?

Now, a being of an immaterial order, which exists independently of space and time, as we have seen, and can, in this logic, be the cause of the existence of space and time, that is, creator of the material universe, as we have also seen in logic previous. Therefore, the same logic indicates that, being the Creator of the material universe necessarily, umbilically, linked to the created being, He can necessarily communicate with it. So much more can there be this communication insofar as some being of the material order has received from its Creator some attribute that makes it similar to its Creator, and the revelation complements us by referring to a being created *in its image and likeness*: a human species.

This attribute of being of the material order, or of being created *in the image and likeness* of its Creator, is clearly revealed in the Consciousness of the human species, a species that, being part of the essence of a finite and temporal being, carries within itself the attributes of being independent of time and space, that is, it participates in the order of the spirit. This is not a question of faith, but of logic and, therefore, beyond experimental science, that is, it is the object of philosophical science, or metaphysics.

It is for this reason that experimental science, not being able to find answers to questions that go beyond its nature, must accept the existence of the same scientific certainty in another order (in the purely logical order) and refer to this other order, existence, or not, of a previous and different cause of matter, previous cause, which gave rise to the Big Bang, initial matter of the universe, or any other theory about the origin of the universe, and for this reason, as it is able to create, it is also able to communicate with its created order, and it is this communication that we call Revelation.

c) **The third logic, if this material being, endowed with a limited spiritual attribute, is able to receive the object of communication from a purely spiritual Being.**

About this capacity, the initial premise is that communication in this kind of knowledge is necessarily limited, due to the limitation of the receiving species whose nature is conditioned by matter. It follows, therefore, an impossibility of the created order to penetrate the absolute or essential nature of its Creator, Being before and beyond matter. Therefore, the created order is incapable of understanding a Being that is beyond any limitation in an absolute way, therefore it is an absolute spirit. In strict language, revealed knowledge exists, but it is of a limited nature, because of the limitation of the human species towards an object of an absolute, unlimited nature.

Thus, the knowledge of an absolute spiritual being must always be limited, and as aptly noted Saint Aquinas about this Being, *we can know more about what He is not than what He actually is.* (Summa Theologiae)

As a result, the only possible way for those seeking an answer to this question is that He Himself, the Immaterial Being, communicate through some form of Revelation, and this grounds faith, and proves how the Revelation, or faith, they complement experimental science and philosophy itself, as well as that faith has with experimental science and philosophy a commitment to complementation, never to discord or denial, a commitment that is, or should be, mutual.

d) The fourth logic is the impossibility of the existence of two or more absolute beings.

This relative purity of knowledge does not inhibit the assertion of a fourth deduction: since this Being is beyond space and time, full and absolute, the possibility of another being with this same nature is excluded, because He, having neither space nor time, is not subject to limit of any kind. In that order, He is everything.

For this reason, from the only relative purity, or analogous, of logic, Consciousness, or human logic, cannot have the same nature as the absolute Being that generated it, just as, for the same reason, there cannot be another absolute being. On the other hand, and in the same way, the existence of spiritual beings can only be admitted as created beings and, therefore, only as attributes of the absolute Being.

For this reason, whether in relation to material beings or in relation to spiritual beings, participation can only be *"in the image and likeness"* of the Absolute Being and never in equality, and in this possible similarity is found the connection that unites logic from reason, or philosophy, to Revelation and theological or biblical explanation, *of the image and similarity* given by the Creator to one, and only one among so many species of His creation in the material order, at least while it is made known to us: the human species. The existence of beings created in the spiritual order remains in the field of theology, faith, or simply the Mystery...

The above confirms what has been said about the limitation of our logic in the knowledge of this original Being and so, just as we crossed the limit between experimental science and philosophy, or metaphysics, we now reach the limit between philosophy and existence, as another level of knowledge, beyond the human dimension and its methods: knowledge through Revelation.

e) The fifth logic is the existence of another level of knowledge: Revelation.

Knowledge by Revelation, or by the theology that studies, interprets, and completes it in its form of human expression, or simply the knowledge that arises from the acceptance of faith, simply means that in addition to experimental science and philosophy there is another method that reveals the truth to us.

Due to this different method of knowledge, the knowledge achieved by Revelation, or by faith, it is necessary to reflect on *the image and similarity* to the Creator, who is the channel or instrument of this communication: the Consciousness, the Spirit or the Soul, *that image and similarity* given to the human species, and for this reason the Revelation is an absolutely valid knowledge, once admitted a creator God who communicates. If there is no such God, or if there is no Consciousness, there would be no validity in the communication. It's evident.

At any time it can be seen that through Consciousness, or thought and other attributes that make up Consciousness, such as, in addition to self-perception, feeling or imagination, acceptance or repulsion, will, we are transported from real form, one could also refer as virtual, but without any subject, object or physical medium, or material, to another time or another space, to the primitive world, or to imagine the world of the future at any time, or anywhere in the universe. This, then, is by no means experimental knowledge.

It is also not a more complete experience, as a purely spiritual knowledge would surely be, because of our limitation by matter.

In any case, this multiple and universal presence of Consciousness in time and space is our participation in the world of the spirit, although it is necessary to return to the alert that, in this world of the spirit, in no way can we be equal to the absolute, unique Being (that one without time and without space), Creator of everything, because, beyond that, we are limited by matter, I repeat.

This ability to have Consciousness present in a world independent of time and space, leads us to conclude that, freed from the limitations to which matter is conditioned, the part of our nature that is freed from matter, time and space, which means the Consciousness and its attributes, or the Spirit, which exists in the human species, will survive in the world of infinity and eternity, because its death, or its destruction, contradicts its immaterial nature, obviously.

Would we then be immortal and eternal and, therefore, the temptation to consider ourselves equal to God returns, or would we be gods? Would there be as many gods as there were human beings freed from matter? Although our analysis lays in the spirit world, therefore, of knowledge through Revelation, followed by theology, there is still a reasoning imposed by the logic of philosophy, which answers this question.

We saw that the existence of this absolute Being, infinite and eternal, excludes the possibility of its duplication, or multiplication – for there cannot exist two or more absolute beings in the same order of being. If there were, both would be limited. This means that, as in this world, we are only *in the image and likeness* of this absolute Being, in this new condition of a freed, immaterial nature, it is only possible for it to participate in its attributes and not to be equal to the absolute Being.

Anticipating a theological thought, as the Bible teaches, that we were made in the image and likeness of the Creator, Saint Paul, theologian, referring to our connection with the Creator, defines that *"in Him we are, we move and exist"*, a principle that, applicable to what we are before the liberation of Consciousness (or the spirit), also applies to what we should be after this liberation (or bodily death).

We are not, therefore, new gods, but we resemble the image of God, his attributes, and in this resemblance are found philosophy and theology, or Revelation, which will tell us, this, Revelation, or theology, for those who believe, how is this Absolute Being and what, or how is this likeness and how it is accomplished here or will be accomplished after liberation.

This is the Mystery, or the answer, of Revelation or of faith. For those who do not believe in Revelation, I mean, in some form of revelation, there remains the absolute mystery, which is carried in different ways, according to each one. Some will remain oblivious, others will seek alternative paths, others, finally, will live in the anguish of the absurdity of remaining in a whole with no explanation for its nature, or in its essential being, carrying a life that actually (consciously or even unconsciously) loses meaning, because everything will end very soon, for what are a few years before existence and the length of time, or eternity?

I know there are those who prefer it that way, trying to construct the absurdity of living in fullness in a world of absolute limitation...

f) In conclusion.

Considering the absolute of eternity and the absolute of infinity as attributes of the Creator and in view of the limitation of the human species, or of matter, the way is open, so that, starting from experimental science, passing through the logic of philosophy, we find in theology, while interpreter of Revelation, or in faith, the way, or the responsibility to answer about the existence of that Absolute Being, unique, eternal beginning, origin and end of all things. I repeat the theologian and apostle Paul, *"in which we are, we move and exist."* (Acts 17:28-30)

The acceptance of this knowledge via Revelation, or Faith, thus becomes absolutely compatible with philosophy and experimental science, each at its own level, but complementing each other and this seems to me absolutely logical and therefore scientific. This logic can be complemented with the logic of Saint Thomas Aquinas, who, in *his Summa Theologiæ*, seeks to rationally prove the existence of this absolute Being, through five ways, or paths of knowledge. [1] (ST, I, q. 2)

---

[1] - *How philosophy, according to St. Aquinas leads us to admit the existence of God.*

In the analysis of this question, two currents develop their method to analyze the question of the existence of God. One tends to the method of logical reasoning that has its origin in the Greek philosopher Aristotle and that had Saint Thomas Aquinas as its maximum exponent in the Christian era. Another current, whose origin is also attributed to a Greek philosopher, Plato, who in the Christian era was "Christianized" by Saint Augustine, and is based above all, as a method, on intuition, feeling and, therefore, on the automatic perception of discovery, or the construction of truth, including the essential truth of God's existence.

As a synthesis of all the aspects that gave rise to scholasticism, the predominant philosophy of the Middle Ages, I initially describe the well-known five paths of knowledge that lead us to God, according to Saint Thomas Aquinas. Then I will make a reference to St. Augustine.

I believe that these very short syntheses provide a minimal and very important knowledge of the roots of the absolute majority of the affirmation trends, or the denial of the existence of God, even in our time, when the greater tendency is to diminish the value of Thomistic logic, to make room for intuition, or feeling, instinct, or passion.

a) The Five Ways of Saint Thomas Aquinas

1st Way - All movement presupposes a first motor. *This is part of the observation that the whole being is in evolution, moves in this direction of evolution. Now, every movement must have an origin that makes it move, because an eternal movement, which loathes limited matter, is inconceivable. Therefore, there must be a Being that is the origin of all movement. This Being is God. Understand this movement as life, evolution, transformation.*

2nd Way – Of the efficient cause. *This way, in a similar way to the previous one, refers, however, to another evident fact: the existence. Everything exists because it had an origin, and it is not possible for this origin to be multiplied to infinity. Consequently, at the origin of all that exists, there must be a cause that was not generated, and which therefore is at the origin, or exists as the origin of all things that exist. This being is God.*

3rd Way – Of the contingent and the necessary. *It also develops by the same reasoning: the essence of everything that exists shows us that everything that exists could not exist. But it must be deduced that it would make no sense for these things to exist if there were not in their origin a being whose essence, whose essential nature is necessary, since if this being did not exist, nothing would exist. God is this necessary Being.*

4th way – About the degrees of perfection. *In the observation or analysis of all things, any conclusion leads us to deduce that something is more or less, better, or worse in relation to something else. But this analysis, insofar as it seeks something more or better, will have to do it successively in relation to something else even more or better. Well, one could not, equally, take this analysis to infinity* *and, for this reason, it becomes imperative to arrive at a being that has absolute perfection. That being of absolute perfection is God.*

5th way - About the organization of the order of things. *Things, all things exist or move in an orderly way, otherwise everything would be in continuous conflict and nothing of the form the universe could exist. Now, in no way could such an order of all things, from the smallest things to the greatest things imaginable, be organized without an intelligent cause, above all things, to order them. This intelligent being who ordered all things is God, like the archer who orders the arrow so that it hits its target. It would never hit the target if it didn't have an archer in command.*

These five ways, exposed by Saint Thomas Aquinas, in his main work Summa Theologiæ, which lead us to admit the necessity of the existence of God, constitute essentially logical reasons, purely intellectual, in the rational line of Aristotelian origin and essential instrument in the philosophy of the scholastic method.

b) St. Augustine's way of intuition

However, it is absolutely valid, for some more, for others less valid, the intuitive perception of the existence of God, the result of human feelings, anxieties, and aspirations, or of human restlessness, referred to by Saint Augustine, in a philosophy representing the Platonic line in the Christian era. This philosophical line is followed by countless philosophers, theologians, sages, and saints, and it is summed up in all human restlessness expressed in a significant phrase: **"You created us Lord, and our hearts will be restless until it finds you, until it does not rest in you".**

I don't see these schools and their multiple differentiations in all times and even in our times, as contrary, but complementary to each other, different paths, but both fruit of human nature, which leads us to the same point: the existence of God.

I conclude, therefore, reaffirming that in the rational world the acceptance of some form of Revelation that complements the limitations imposed by our material nature is a rational attitude, therefore scientific in its order, and, so, it also constitutes a response to the absurdity of living without an answer to the essentials of existence, as I have just recorded.

1.3. Where are we going?

Having posed the question of the origin of the universe, before analyzing the origin of life in the universe and of Consciousness in the human species, mentioned several times, I still want to reflect on where we are going, the human species, and where the universe is going, questions to which, due to their essentiality, I will return at the end of this book.

In relation to the human species, the answer can only come, and does, from its nature, that is, from its essential identity. Apparently, everything ends in death, but, as we've already seen, only apparently.

This is because the human species, unlike the rest of the universe, was given a share in the attributes of the spirit, through Consciousness, or in Consciousness...

a) About the dimension of Consciousness

- If the Consciousness, despite the limitation imposed by matter, frees us from time and space, allowing us to transport ourselves anywhere and at any time... and, especially, also inside ourselves in self-perception;

- if the liberation of space and time is in the essential nature of the spirit, which is immaterial by nature and, therefore, eternal, logic, considered the nature of the human species, this makes us conclude that the death of matter does not mean the death of the spirit and which, consequently, does not even characterize the end of the individual of the human species;

- it means, on the contrary, its liberation from space and time, which condition and limit their Consciousness, or the spirit in which the person, and the species, participate;

b) About the release of Consciousness

- The Consciousness, therefore, which allows the human species to take part in the nature of the spirit, does not cease to exist when freeing itself from the limiting matter, which leads to the conclusion that, somehow, this release will make the spirit return to the participation of its origin, and this means returning to the dimension of the nature of *whom it is image and likeness*, that is, of its Creator, from where it came.

The Mystery of this return, however, takes us again, from this logical equation to the answer of theology or of Revelation, as well defined by St. Paul, already quoted, when he says that *"in God we are, we move and exist."*

This gives us security, through faith, complemented by the logic of philosophy, and independently of experimental science, that death frees us from time and space, to make us return to the world of spirit, of infinity and eternity.

There remains the destiny of the body, which, in the same way, will only have an answer in the field of theology, if it has one, since the Bible only brings us some specific examples of resurrection of bodies, like that of Lazarus, or of the son of the widow of Nain, or of the daughter of Jairus, by Jesus' miracle, even Jesus' own Resurrection three days after His death on the cross and, according to tradition, theology and the official doctrine of the Catholic Church, the resurrection of the body of the Virgin Mary, Mother of Jesus, raised to heaven.

But it is necessary to realize that in these examples there is an essential difference: Lazarus, or the son of the widow of Nain, or the daughter of Jairus were resurrected to this world, while Jesus and Mary were resurrected to the other world, but surely with a transformation of matter; what transformation, or how... the Mystery remains...

Anyway, these cases are exceptional, not applicable to all species, and for this reason the Mystery continues, in the field of science fiction to experimental science, or from speculation to philosophy, leaving the Mystery in the field of Revelation, or of faith.

However, considering this, it is imperative that somehow the body, the matter, must join the spirit, so that the person survives in eternity as a human person, not as a spirit, angel, or any other form of pure spirit, and does not cease to exist as an individual, or assumes another identity no longer being the same person, if the survival were only of Consciousness, or of its spiritual dimension.

c) *About the fate of the human species.*

In addition to this individual destiny, in matter and in spirit, the question of survival, or the destiny of the species, remains. One can ramble away from science and revelation, and imagine that the human species could be extinguished by the successive decrease in the generation of new individuals, or by some catastrophe like the one that would have extinguished the dinosaurs, or by some mutation of the species that would end, in some way, to put an end to its reproductive capacity, or through the work of the Being that created it, to return as a species to the Creator, stripping itself of matter by the prevalence of spirit, as the maggot transforms itself when leaving the cocoon.

Everything is allowed to the imagination, including the hypothesis that we are still in the prehistory of the human species, and that we would have been made to inhabit the universe...

However, if in relation to the individual end, the Mystery may have been unveiled, at least in part...

- if in relation to the human species, the imagination can be satisfied in rambling, it still remains in the same, or greater Mystery, the destiny of the universe...

1.4. Where does the universe go?

When talking about the fate of the universe, we return again to the analysis of material being, therefore, to the world of space and time, of quantity and form, and we return, thus, to the field of experimental science.

Several hypotheses are put forward by experimental science, which makes use of telescopic observation, mathematics or, also, imagination. Scientists at all levels of science, have pursued these attempts, complementary to each other, similar or contrary, since the philosophers, as did, for example, Aristotle and St. Thomas Aquinas, as the greatest physicists or astronomers, for example, the French priest and astronomer Lemaitre and physicist Einstein, or even other scientists more popularized as outstanding communicators such as Isaac Azimov or Stephen Hawkins, cited above.

Apart from that, I want to refer to the theory of the anthropologist and theologian Teilhard de Chardin, who, more than mathematics and telescopes, drew on anthropological research and on the inspiration of theology, from the origin of the universe to the origin of man and its destiny.

From them, or from others less well known, the most diverse theories emerged, some complementary, others parallel and some, as I said, with contradictory nuances in relation to the others. I make a quick reference to the best known, developed by physicists and astronomers:

a) The theory of the expanding universe

From the initial explosion, the Big Bang, pushing the matter generated through certain energy – dark energy, to the extremes. When this energy completely loses its strength, everything will disintegrate.

But the questions remain: where did the original Big Bang energy come from and what mystery will come to exist from this disintegrated universe?

b) A complementary theory

The answer imagined by a theory that is somewhat complementary to the previous one, states that the same energy, instead of disintegrating the universe, will end up concentrating matter to the point of remaking the initial atom, giving rise to a new Big Bang, and like that indefinitely, matter expanding and concentrating. But is this an answer, or is it simply a way of pushing the mystery further?

c) Black holes theory

At the same time, there is the theory of black holes. This theory states that black holes, this antimatter, will eventually "swallow" all the matter in the universe, continuing the process in which the larger black holes will swallow the smaller ones giving rise to a new universe, where matter will lose all its identity, become immobile, dark, let's say, leaving only an image of our universe, perhaps a shadow, the antimatter...

But again, the question remains: does this antimatter exist? If so, what is it? Will it be eternal? Is it infinite or will it not exist? Is it nothing? But does nothing exist? (This is not a play on words or a toy, but an essential question to answer... the Mystery...)

d) The theory of anthropic energy

The theory of anthropic energy assumes that energy will equalize all existing bodies in the universe, generating a cold and totally immobile universe, thus ending all current physical or astronomical phenomena, making any form of life unfeasible, in addition to rendering unfeasible time and space. The question that keeps coming back is whether this totally cold or totally immobile universe, in front of our concepts, is even thinkable.

Would it be minimally scientific to admit a world that is not even thinkable, or would it be just a way of, once again, pushing, or dodging, the Mystery?

e) The theory of the end of time

The end-time theory complements that of anthropic energy and states that, as our universe is finite in nature, its expansion, admitted in other theories, could only expand it to its limits and at this moment, the universe will freeze, and immobile, will put an end the existence of time, simply passing, the universe, not to have any sense within our concepts.

But the same question remains: would this immobile universe, which would overturn all the concepts that govern the present universe, solve the Mystery?

In addition to these theories, the most in vogue, based on mathematical calculations, telescopic observations and other instruments, with greater or lesser use of imagination, mathematics or logarithms, I want to comment on the anthropological theory I referred to by Teilhard de Chardin.

f) Teilhard de Chardin's theory

With a much broader vision, a synthesis of anthropology, a cosmological vision, with the illumination of Revelation, or of theology, Jesuit priest that he was, Teilhard dedicated his life to research in areas that happen to be rich for this finality, in Africa and Asia.

Chardin conceives of the whole process as an arrow that, launched from a beginning, which he calls the Alpha point, hits a certain final object, which he calls the Omega point – the beginning and the end. The Alpha and the Omega constitute the same Creator Being, as nothing exists without the cause of its existence and without an objective to arrive, initial cause and final cause, principles that Thomas Aquinas defines and defends in his complete treatise on philosophy-theology, already referred to, the *Summa Theologiae*.

In line with this principle, the path of the arrow travels in a way, from simple to complex, as demonstrated by a succession of researches carried out over more than 20 years, especially in China, having as high points of this path the emergence of life and the human species, moments when, once prepared the raw material in this evolution from the simple to the complex (compare a primitive organism with the complexity of the human organism), new interventions by the Creator gave rise to Life and Man in the universe.

This path of complexification, starting from the Absolute perfection, the Creator Alpha, in this process of continuous improvement towards dematerialization, or spiritualization, will reach again the Absolute, the same Creator God, from which it originated.

A reference here, again to Saint Thomas Aquinas, who in the process of creation finds the essential reference of the very nature of God. Although Aquinas has devoted himself in depth to his studies on the nature of God - they are well known in his greatest work, the Summa Theologiae, the 5 ways through which he seeks to prove the existence of God - he affirms the principle that love has itself the tendency to spread – *Amor diffusivum sui*. (ST, I, q. 5)

From this principle of the great Doctor of the Church, it can be safely deduced that, since the process of creation is a work, an expansion or an overflow of God, the essence of love (and this also explains the love present in the process of Redemption, as we shall see), its source, consequently the destiny of the human species, as indeed of all creation, is the return to Love, as we shall consider again in the final part.

It is recorded that the Catholic Church, or its theologians, took a long time to analyze and accept Teilhard de Chardin's work, which was finally published, in its essence and with its endorsement, by the Church, in the book "The Human Phenomenon", edited in 1948, two years after his death.

2) The Answers of Theology, or Faith

2.1. The assumptions of the answers

At all times, all peoples have always looked to faith for the answer to their questions, or to religion for an answer to their needs. In fact, this is a challenge that transcends the discoveries, theories, and hypotheses of researchers in the experimental sciences, mathematics, or the lenses of their space-penetrating telescopes, with distances increasingly measured in light years, or microscopes that penetrate the infinitesimal particles of matter.

But their instruments did not manage to penetrate time, be it the past or the future, except in fiction...

The essential truth is that there will be no answers given by experimental or experiential instruments or methodologies about realities that transcend the quantitative, space and time and, therefore, are situated in the infinite and eternity never experienced by these methods.

Ancient peoples in their sacred writings that codified their faith, their religions, or their morals, in general also sought answers about the origin of the universe. I have already mentioned some of these answers.

Referring now to this fact, the origin of the universe, I stick to the narrative of Genesis, the initial text of the Bible, the sacred book of the Hebrew people, in the version referred to at the beginning, a book that inspires, in addition to religion itself of this people, answers also to the Christian religions, those which, in the diversity of their cultures, or their beliefs, had their origin in it.

Therefore, I stick to Genesis, the first book of the Bible in this version, when reflecting on the origin of the Universe, as I stick to the same version when referring to other texts from this or other books of the Bible in other questions where I should look for the logic between experimental science, philosophy, theology and Revelation, or faith.

Finally, I stick to it, to the Bible in this version, because I find in it the foundations of my faith, although other versions of the Bible or other sacred books seek and propose the same goals, albeit through alternative channels or forms.

I make this choice above all because I identify in it an absolutely significant logic between it and the various questions, or Mysteries, object of the reflections in this book, a logic that, in my opinion, I consider adequate and sufficient.

Regardless of my choice, I think that in other sacred books the same or similar logic can be found with the respective cultures, by whoever comes to look for fundamentals in them, being willing to research them. But this search in other sacred writings would, at that moment, go beyond the dimensions and purpose of this book. ²

² *On the issue of the existence of a wide variety of translations of the BIBLE, Sacred Book of Christianity and Judaism, which justify the clarification I made to page 4 of "Mystery of Everything", journalist, writer and researcher Valério Azevedo, author of the presentation of this book, he left the following contribution: "In human history there are numerous collections of sacred books, conceived, since the beginning and throughout the world, by peoples and cultures to affirm their beliefs. This considered, none is greater than the others, as in general, it can be said that everyone seeks the same truth. Among many, several can be cited from older or remote times, such as the KORAN, in Islam, the AVESTA, from Zoroastrianism, the VEDAS, from Hinduism, the CONFUCIUS DIALOGUES, from Confucianism, THE BOOK OF TAO,*

from Taoism, or from times moderns such as Alan Kardec's BOOK OF SPIRITS, THE BOOK OF THE NEW WORLD of Mormons, and others...

In addition to these, there is the millenary set of parchments, tablets and papyrus scrolls, which served, and continue to serve as research sources for all those interested in better understanding this rich repertoire of sources, also to understand the Mystery of the Universe, the Mystery of Man and the Mystery of Redemption, as we do in this book, or any other matters of biblical interest.

The various versions of the biblical text approved by the Catholic Church are generally based on the Latin Vulgate, a text translated from Greek, Hebrew, and Aramaic originals, still considered the variety of texts previously existing in Latin, under the name of Vetus Latina, or Latin version. The Vulgate was translated or compiled by Saint Geronimo at the behest of the Bishop of Rome Saint Damasus,

Pope, at the time of Emperor Constantine the Great, who had it promulgated in Latin so that it could be read and understood throughout the land.

Translated into French by the monks of the monastery of Maretsou, Belgium at the beginning of the last century, as the Author of this book informs, the edition was translated into Portuguese by the Catholic Biblical Center and printed by Editora Ave Maria in São Paulo, in its 52nd edition, in the year of 1957 of the Grace of Our Lord."

Genesis attributes to a spiritual and creator Being, as previously mentioned in this book, the only and absolute God, the creation of the universe, life and man, a work that, according to the knowledge of the time and in the language of his time, took place in six days, and then God, on the seventh day, rested.

Today theologians agree that the biblical narrative is an evident allegory of creation, as a sequential act of the Creator – evolving each day as corresponding to a step into a given evolutionary process. The same symbolic meaning is attributed to the seventh day, when, according to the Book of Genesis, God rested and handed over his work to man's care – the supreme crown of creation. In fact, the seventh-day narrative assigns to man co-responsibility for the continuity of Creation, that is, for the preservation and improvement of the created work.

The Bible, the sacred book of Jews and Christians, does not originate from scientific research and does not intend to be a scientific narrative, as we understand science today. As a book of faith, it originates from traditions, or from the Creator's revelation through inspired men, prophets, scribes, and leaders of the Jewish people (and then Christian leaders), who throughout the centuries have written and rewrote it.

Therefore, it is understandable that, being a work built by dozens and dozens of generations, it has been over the centuries, added to innumerable cosmogonies, elucubrations, theses, theories, mythology, mysticism, cultures or political conjunctures, which often may have had their contributions included more in line with the editors' vision than with the success of divine inspiration.

Thus, this narrative is not aimed at experimental science, but given its nature of revealed knowledge, it is situated in a different world, supernatural or transcendent, above or beyond matter conditioned to the measures of space and time.

There is, however, a greater connection between the revelation contained in the Bible and the logic pertaining to philosophy, because both are independent of experimentation, and despite the difference in method, both logic and Revelation cannot accept the absurd. Thus, when Revelation confronts us with a Mystery, the revelation of that Mystery cannot oppose logic. I mean that although the Mystery may not be explained immediately, or may never be explained, its existence does not jeopardize the principles of logic.

This is a question to be pondered by those who, in the search for truth, seek knowledge beyond purely experimental science. This is what the theologian, already quoted several times, Thomas Aquinas, father of scholastic philosophy, teaches when he affirms *Praestet fides suplementum, sensuum defectui* (Sto Tomaz: Hymn to the Blessed Sacrament-Pange language.), which means that faith must serve as a supplement where the senses no longer reach. Perfect complementation between experimental Science, Philosophy, Theology and Revelation, or faith.

Understanding the revelation or faith contained in the biblical narrative in this way, it must be seen that these are not opposed to experimental sciences or philosophy. On the contrary, within its method, its own way of narrating, in each of them the same truth can be found, both of which can even serve as a light that illuminates the path in this search.

2.2. The origin of the universe

Assuming that there is no effect without a cause, and this is fundamental in any science, in any logic or in any level of knowledge, Revelation affirms its first truth: the creation of the Universe, therefore, the existence of a creative Being.

Consequently, and for this reason, the creative cause existed before the universe, which leads us to admit that this cause is endowed with a nature that goes beyond space and time, and we have already seen, realities that did not exist before the existence of the universe, as we understand it, obviously.

Existing before space and time, it also follows that this cause is eternal and infinite, at least according to our concepts of space and time, infinity, and eternity, therefore, as we have also seen, it can be the creating cause of the universe and, therefore, from space and time. The issue must be understood beyond the literal revelation of the Bible.

The text does not affirm creation as a single and complete act, but as something that takes place in successive stages, represented by the seven days of creation.

Once the work of creation was finished, the Bible reveals that the universe was handed over to the masterpiece of its Creator, Man, the human species, to whom the Creator endowed a portion of himself: Consciousness, the spiritual attribute and with it the ability to create, the ability to penetrate beyond space and time, while conditioning it to a material dimension, uniting it to a material being.

To this capacity constituted in *His image and likeness*, the Consciousness given to the human species, the Creator wanted to hand over the responsibility of taking care of his work, using it and completing it. This is how the biblical text communicates to us: *"God created humankind in His image, created them in the image of God, created them male and female. God blessed them, and God said to them, "Be fruitful and multiply, and fill the earth and subdue it; and have dominion over the fish of the sea and over the birds of the air and over every living thing that moves upon the earth."* (Genesis 1:27-28)

It is impossible, therefore, not to relate Genesis to the concepts of the experimental sciences of evolution. Evolution only enhances the work of the Creator, uniting it deeply to Him throughout all times and to the Revelation of the work of creation, which only illuminates or confirms the search for science and its conclusions.

This means that the Creator continues to complete His creative work, through man, of the species made in His image and likeness. This truth gives man an unimaginable dimension and dignity and this will be analyzed later.

In this process of evolution, three essential stages can be identified: the creation of the material universe in the first, second and third days; the creation of aquatic life on the fourth day; on the fifth day, terrestrial life, animals, and greenery to feed them. Only on the sixth day did God create man and give him the universe with everything it contains, so that he could take care of it and use it to satisfy his needs, this way completing and giving a new meaning to His Creation, in the human species.

Can there be a more dignified vision for the human species than this?

Revealed the existence of the cause of Creation - the Creator Being who answers all the questions left unanswered by experimental Science - the question remains about what we can know about this Creator, who made the universe and all the things in it, including the human species.

It is evident that this answer will not be reached by the experimental sciences, or even by philosophy, because when reflecting on the nature of the Creator, we are beyond space and time, beyond any experienceable form and even beyond human logic, for lack of knowledge of its premises or its object. With regard to its essential nature, the answer will only exist insofar as this God reveals Himself to His creature and, therefore, we will have to remain in the field of Revelation, it means, the interpretation or inspiration of Theology, or adherence to faith.

Let us penetrate this Mystery.

2.3. About the nature of the Creator Being

The Bible, through Saint John, apostle of Jesus, the Redeemer, states in his Gospel that *in the beginning was the Word, and the Word was with God and the Word was God*, and then completes: *"in the beginning God created heaven and the earth, all things were made through Him and without Him nothing that was made was made."* (John 1:1-3)

Considering that with these words the Apostle begins the narration of the life of Jesus, we can interpret that the Word, the power of God the Creator, reveals in Jesus a second divine person, who would redeem the masterpiece of creation, the human species.

And I add more...whose eternal divine wisdom knew that in this masterpiece, He would grant something in *His image and likeness*, Consciousness, which means He would grant Liberty and with Liberty, the merit of all the good that was done, but there would also be the hypothesis of the practice of evil, a hypothesis that would be confirmed from the disobedience, or initial fall of the first couple, according to the biblical narrative and repeating itself through their descendants.

Following the same Gospel, as in fact in the Gospels of the other evangelists confirmed by the Catholic Church, a third divine person, the Holy Spirit, is revealed several times, with whom the nature of God the Creator is completed.

One God in three persons: Father, Son and Holy Spirit, to whom the functions of Creation, Redemption and Sanctification, respectively, are associated in theology. It also reveals, in the presence of the Holy Spirit, a form of intimate union, or the Love of God for His creature – not equal to Him but created in *His image and likeness*. This Love often reveals itself as one of the most significant moments of this revelation coming the human generation from the second Person, Jesus, the Redeemer, in the womb of Mary.

When young Mary showing her fear before the Angel who announced to her that she would be the mother of a boy whom she would name Jesus, full of fear asked her: *"Mary said to the angel, 'How can this be, since I am a virgin?' The angel said to her, 'The Holy Spirit will come upon you, and the power of the Most High will overshadow you; therefore the child to be born will be holy; He will be called Son of God.'"* (Luke 1:34-35).

The narrative is confirmed by the evangelist Matthew: *"Now the birth of Jesus the Messiah took place in this way. When his mother Mary had been engaged to Joseph, but before they lived together, she was found to be with child from the Holy Spirit. Her husband Joseph, being a righteous man and unwilling to expose her to public disgrace, planned to dismiss her quietly. But just when he had resolved to do this, an angel of the Lord appeared to him in a dream and said, 'Joseph, son of David, do not be afraid to take Mary as your wife, for the child conceived in her is from the Holy Spirit.'"* (Matthew 1:18-20).

This presence of the Holy Spirit is revealed in many other moments of the Gospels, as in the Baptism of Jesus, or in the meetings of the apostles after the death of Jesus, closed in the Upper Room for fear of the Jews, giving them charisma and courage to spread the Message of Redemption to the world.

In a logical reasoning, these narratives make absolute sense, especially in relation to the nature of Jesus and the virginity of Mary, since Jesus, being a son of God, would keep His divine nature, thus not justifying the interference of a biological father, but only, giving it the full human dimension through the incarnation through the womb of a woman, so that the Redeemer, in this way, together with his divine condition, could assume his human condition in its fullness. Mary offered this breast: *"Here am I, the servant of the Lord; let it be with me according to your word."* (Luke 1:38)

I believe that this set of narratives, considering the life and Message of Jesus, including His death and resurrection and the continuous intervention of the Holy Spirit, opens the door to the revelation of the nature of God through a second and a third Person, nature which, although it does not explain the essence of the Mystery of the existence of a single God, unique but existing in three distinct persons forming a divine Trinity, it explains the cause that existed before space and time, that created space and time, or be the universe and everything in it.

In this Trinity it is understood the permanent and absolute presence of God as Creator, Redeemer and Sanctifier of His creation and, I repeat, work to which the human species is essentially linked, and in a sensitive way, above all, in the motherhood of Mary from which all creation and its history were represented, including the original Consciousness of the human species, whose Freedom, mistakenly used, gave rise to the entire project of the Redemption.

I know this is a realization of immense size, the meaning of the human species, the Universe, and its Creator. But it is in this dimension that divine and human things meet and take on meaning. This is also the foundation of the essentials of faith, of my faith.

No. **All things are not explained, but all divine and human things become logical**, and impose themselves on this limit, since the Consciousness, the bond of this union of the Creator with His creature is limited, belongs to two realities, the material, space and time, and that of spirit, infinity and eternity.

In this way, it gives us the path, which allows us, in addition to accepting, to understand the logic of the existence of the inexplicable in science, in Revelation or in faith, about where we are, in what reality we live. I give as an example, perhaps allowing a better understanding of the theology of the Apostle Paul, when referring to God stating that *in Him, we are, live and exist.* (Acts 17: 28) or St. Thomas' most often quoted statement, which teaches that faith serves as a complement when the senses can no longer understand.

But ultimately, due to the dimension of the Mystery, or the mysteries contained in this whole process, logic is on the path of acceptance of Revelation and Faith, or will it remain in the absurdity of the absolute lack of understanding of the essentials of all things, or, worse, not even knowing that the essential of all things exists.

2.4. About the origin of life

Before analyzing the mystery of man, or the human species, the crown of the work of creation, it seems necessary to reflect on the origin of life, in a dead universe before life existed, repeating, in other words, the assumption of the evolutionary creation of the universe.

Let's start by considering that the Bible itself, in the same book of Genesis, informs: *"the earth was a formless void and darkness covered the face of the deep, while a wind from God swept over the face of the waters."* (Genesis1:2) This means it had no life. And, after explaining about the evolution of the universe, having separated the earth among itself, that earth that was once formless and empty, from the waters and the firmament, and having the light dispelled the darkness, the Book of Genesis goes on to report the creation of the "green" of plant life, to cover the surface of the earth and feed the life that was to come, all this on the first, second and third day.

On the fourth day the Creator made the aquatic animals, Genesis continues, and commanded them to populate the waters with their multiple species. On the fifth day, once the land was separated from the waters, the land covered with vegetation, the luminaries created in the firmament so that there might be light, God created the land animals, and ordered them to populate the land as well, again in the variety of their species. Thus, prepared the entire universe, prepared and ornamented the earth, on the sixth day, Genesis reports the coronation of God's creative work, with the creation of man, *made in His image and likeness.* (Genesis 1:26)

But so far, I reaffirm, it is impossible not to see in this narrative the successive evolution of the Universe and Life on Earth and it is even strange that Science has taken so many centuries, or millennia, to discover the evolution of the work of creation and all order, all the greatness and all the power, or the wisdom of the Creator, while the same delay had, or still has, many religious groups, to understand and accept it in its greatness, or in its fullness.

Thus, there is no doubt that the essential fact of the creation and evolution of the universe and life until the arrival of the human species is clearly stated or taught in Genesis.

It is up to experimental Science to define how this evolution took place, since the Bible, not being a book of Science, but of faith, as already said, or of Revelation of the essential truth, creation and its Creator, cannot be taken as science book, nor would it be able to pre-empt science by thousands of years.

Thus, the biblical description, revealed the essential truth of God the Creator and of Creation, rightly left it to men, to science, to discover the way in which evolution, the scientific truth, took place. The Bible was written by the prophets or by the sages of the time, according to the knowledge of the time and the objectives of the Revelation: the presence of God in the process, as Creator of the Universe, earth, life and man, as the supreme work of creation, as has also been said.

Or is there still another species, after the human species, with man still having to reach the full dimension of the spirit – a new step in evolution, or a Mystery linked to his role or destiny in the divine project of Creation? The answer is according to the choice of each one...

If according to experimental science, philosophy, or theology, there is no doubt about the creative evolution of life, whether plant, animal, or man as belonging to this species (animal), when asking about the origin of life, where it came from, there begins the challenge, first to science, then to theology or faith.

Attempts to respond to this challenge by science, since the most ancient times, have taken two paths in particular:

*a) **The first, of spontaneous generation.***

The first attempt at an answer, from the most ancient times, evolving to the most recent interpretations, which try to explain the origin of life through chemical combinations of primary elements. These attempts do not seem to realize that the transition from non-life to life is not explained by a simple evolution, but necessarily, when dealing with the existence of a new nature, it literally means the existence of some form of creation.

b) The second, of panspermia.

The second attempt, generically used the term panspermia, believes that life would have come from other celestial bodies through space. However much one tries explanations through spontaneous generation or through panspermia, more or less experimental or perhaps more or less imaginative, I don't see how they can be a valid explanation for the question.

c) About the inconsistency or the condition of these hypotheses.

In fact, in relation to the first hypothesis of spontaneous generation, the question of cause must always remain. Whether or not there is some form of generation through primary elements, it remains to be explained where this generative power of these elements that transform non-life into life comes from, even if primitive... which would be permissible, I believe, if the existence of an imprinted potentiality by the Creator was admitted, in what I would call pre-life.

Otherwise, there will always be this leap that goes beyond mutation, as I said, from non-life to life. In fact, it is necessary to understand that mutation, or evolution, is one thing; not life to life, it's something else.

Regarding the second hypothesis of panspermia in its various aspects, it is necessary to realize that, if the coming of life from other celestial bodies is proven, there will not be an answer about the origin of life, but only an explanation of where life came on Earth, perhaps from a certain point, leaving the possibility that it also came from other points.

This hypothesis simply shifts the search for the answer to other celestial bodies, or to other spaces, but the question remains: how did life, transported to Earth from these other spaces, originate?

d) From the view of theology

In the absence of an answer from the experimental sciences, or from the logic of philosophy, which does nothing more than return to the question about the search for the cause, it is left to appeal to theology, or to Revelation. In theology's response to the origin of life, a debate persists among those who defend direct divine intervention not only in Creation, but in each stage of Creation, as we have seen in the case of the passage from non-life to life, denying, therefore, that the created being has in itself the gift or the capacity to evolve from one of these two stages to the other. This is a thread from the debate.

But there are those who, safeguarding the divine presence at the origin and his presence as a result, in the process of evolution, having given the creature the gift of evolving, propose that, respecting this, science should answer about where life began, in what way did it start, or if it only started on our Planet or in other bodies in the universe.

In the wake of these options, other questions arise, as we shall see. In general, the question of these hypotheses only involves a dimension of greatness, or the dimension of the process of creation itself and, in it, of the human species, as I express, without any repair, in my book **Participation and Solidarity**, which is not a book of theology and can be cited at most, a little beyond the experimental sciences, reaching some philosophical dimension.

In this dimension, the book refers to the greatness, or dimension, of human Consciousness and its consequent responsibility, in case it is claimed to be the only consciousness of the universe, and how without this dimension, Consciousness would become diminished, in the case of restricting its dimension to Planet Earth, assuming, by assumption, the existence of other conscious systems in the universe.

The book literally says referring to human consciousness: *"or is it not the only human Consciousness in the Universe? Those who have a tiny vision of man* (because he is the only responsible for our planet, Earth) *will immediately say that he is not. Those who are afraid to admit such a dimension in man* (because their responsibility also extends to the universe) *will also tend to say no. The others, those who remain in creative doubt, will continue in the search and will assume, in revelation or in faith, its complementation, as taught by Saint Thomas Aquinas: Praestet fides suplementum sensuum defectui"*.

So, serve faith as supplement when senses fail. At this moment, therefore, the responsibility for the response returns – the supplement of which St. Thomas speaks, to the world of Revelation and faith or, eventually, to the interpretation of theology.

This considered, it seems that Carl Sagan, when stating that if we were the only species to inhabit the universe, the universe would be a huge waste, he did not realize that the issue is not that simple, disregarding or despising the hypothesis that this *created species in the image and likeness of God*, at this moment it could only be in the prehistory of its own history, or the history of creation and that, for this reason, at this moment in this history, the human species could be just the seed of what is reserved for it, to be, one day, in the universe, in thousands or millions of years because of the acceleration of changes, considered the divine design of Creation...

PART 2 - THE MYSTERY OF THE HUMAN SPECIES

1. On the Origin of the Human Species

1.1. What do the experimental sciences and logic say?

If doubts remain regarding the origin of life and if there is no doubt from the point of view of experimental sciences, philosophy and theology about the process of evolution, the doubt appears when, suddenly, in this evolutionary process a diverse, different being appears among the others, not in form, but in nature, a being that participates in matter, that participates in life, but who is endowed with a different attribute that essentially differentiates him from his ancestors and from the evolutionary process that took place up to that moment.

Although this different being, as matter and as a living being, brings dependence on time and space, finitude and limit, this new attribute frees it, elevates it beyond the conditioning factors of matter, partially frees it from time and space, partially, because his participation in matter limits him in this liberation.

Now, the liberation from the constraints of space and time, as we have seen, means to be, or to participate, in another order of being, of another nature: the spirit, or in other words, it means that this strange being in the process of evolution, the being human, participates simultaneously in a double nature: one of matter and other of the spirit. That's why it becomes a being apart in creation and in its evolutionary process.

Of matter, it can be said that it constitutes an organism, the most complex and ordered, therefore, the most complete of beings in evolution. This complexity allows it to perform a wide variety of functions, which is only possible because, being the most complex, it is also the most organized, as greater complexity presupposes greater organization, otherwise it will turn into chaos.

However, no matter how great the organization of organic complexity, there is no way to explain, only through the resulting organicity, the liberation of space and time since complexity is still material. If there is this liberation of space and time, it is imperative to admit that this organic complexity, in some way, is shaped by a quality of another nature: the spirit.

This new nature, let's say third nature, resulting from the presence, or from the union of matter and spirit existing in unity, constitutes human consciousness itself, with its variety of functions that neither matter, even though it lives, nor pure spirit itself, could explain.

Consciousness, Spirit, Soul, it makes no difference the name given to it.

The functions that are independent of time and space, attributes of the spirit, in the human species become dependent on matter, are conditioned to the limits relative to space and time and thus characterize the Consciousness or the very identity of this evolved but limited species.

Thus, although self-perception, thought, innovation, abstract knowledge, human feeling itself, are specific functions, exclusive to Consciousness, although these attributes can take us, or transfer us instantly, to any place, or to any time, there is a limitation on this transfer. For example, not knowing everything, thought, or feeling or any other attribute of Consciousness, cannot make us present simultaneously everywhere or at all times, or by themselves reveal to us the nature of the facts or relationships that we make present. The body, which limits us, does not participate in the dimension of consciousness and, therefore, we are definitely dependent on matter.

These and other limitations of Consciousness, apart from our participation on the spirit, are imposed on us, because, by this conditioning to matter, I repeat and try to explain it better. This is the reason why the material nature of the organism, which contains or is substantially united with the spiritual dimension, exerts or conditions everything, including the spirit, to all the limitations inherent in matter.

Assuming, therefore, that the less does not explain the more, unless the more is already embedded in the less, or more simply, assuming that there is no effect without a cause, it is logical and rational to admit that if somehow the spirit was infused by an external action, from outside the material being, although alive and evolved, although admirably organized (the complexity ordered in organs).

It was this complex and ordered organism that gave the human species the conditions to receive the spirit - the Consciousness, and to functionate with participation in this other order of being, or perception and other attributes of a spiritual nature, but it would not be able to generate itself, because they are two ways of being of different natures and this is evident.

In view of the above, it is necessary to consider, although, as a result, only actions independent of space and time, even if limited, are specifically human, such as thought and the other attributes of Consciousness already mentioned, which allow us to be present in anytime and anywhere; the self-awareness that identifies us as a unity existing before ourselves; the creativity that allows us to imagine new and non-existent things; the ability, finally, to create, to think or to do new and different things, presuppositions of freedom. Perceiving, self-perceiving, thinking, creating, imagining, these are typical and exclusively human functions, resulting from our spiritual nature, although limited by our material nature at the same time.

Therefore, it is a mistake to speak, unless it is only in an analogous way, in artificial intelligence or animal consciousness, when these functions, in an analogous way, I repeat, are shared by beings of an exclusively animal nature, or only material, as the so-called intelligent machines, also called artificial intelligence.

However, this type of animal or artificial functions, although analogous, can in no way be confused with conscious human actions, as they depend entirely (it will never be possible for the dog to tell the story of its ancestors, or that it wishes to live in Australia, being in Brazil) of space and time and, therefore, not being part of the same nature as human consciousness, spiritual, or human functions, or simply the human species, the functions performed by the human species. This distinction is fundamental.

For all these reasons, considering that a cause of equivalent or greater than the effect is required for every purpose, it is necessary for the creation of the human species to participate in a spiritual dimension, which, in addition to the process of evolution, has also been an object of a creative act of one who already possesses this attribute, the spiritual attribute.

Only an external and pre-existing power can explain the appearance in the evolutionary process of a being endowed with this double nature, that is, material, living nature, the result of a long process of evolution and spiritual nature, made into a single identity or, that is, a new nature, a third nature unlike anything else produced by evolution.

This being simultaneously material and spiritual, made the new species unique in the creative work, unique and different from all Creation (at least from creation as known). It is at this point that the experimental sciences and philosophy itself are surpassed, sending the remaining Mystery to the domains of theology, Revelation or faith. Unless one chooses to renounce all understanding of the essentials, which would mean one chooses once again the absurdity of not understanding...

I stick to the biblical narrative about the creation of the human species, about this matter-spirit junction, to see how far Revelation and faith, or theology, can help us to understand, or to admit, the Mystery.

1.2. The Narrative of the Bible and the Interpretation of Theology

a) Prior understanding.

As seen in the first part of this book on the creation of the universe, the biblical narrative should not be considered strictly a scientific narrative, as the Bible is a book that, drawing on the knowledge in vogue at the time, or at the times in which each part was written, aimed, under divine inspiration, to teach the essential truth about the creation of the universe, with everything that exists in it, in successive days or stages, including the creation of man, as well as everything related to its Creator. But it was essentially neither a history book, nor an anthropology book, or any experimental science book.

In the previous item, it was seen that, from an experimental point of view or even from the logic of philosophy, it is not explained, through simple evolution, attributes of the nature of the spiritual being in which this species, the human species, participates, it means, the participation of something beyond space and time.

The biblical narrative, therefore, just as it cannot be read as a book of science, but as a book of divine Revelation on matters of an essential nature, it must, with regard to historical or material aspects, be understood from the perspective of knowledge, or from the human realities of the times in which they were written.

It is for lack of this perception that the Bible is often misinterpreted, when in its material dimensions - history, customs, language, etc. It is seen not in the light of these realities of the time it was written, but in the light of knowledge, doctrine, or modern behaviors of today.

Therefore, to be well understood or interpreted in the biblical narrative, its revealed contents, its essential or transcendental message, from the form of that narrative, as it was said or expressed according to knowledge, customs, values, etc., must be properly distinguished of the different times in which it was written, I repeat once more, because this distinction is absolutely essential.

I think and I believe that this is a general thought of those who seriously consider the issue, that fidelity to the knowledge and cultures of the time in which the facts were narrated, only increases the narrative's credibility and validity, allowing for a better understanding of the from the scenarios described, its essential meaning, which is revealed.

If this perception is considered and, in this case, especially considered this distinction, the biblical narrative can be better interpreted, such as:

- whether the creation of the human species happened from a couple or from an entire species that evolved;

- if the prevalence of man over woman, which is present not only in Genesis but in general, in other books of the Bible, means a sexist concept, or if it only becomes reprehensible if seen in the light of current values;

- if, as can be seen in the episode of Cain's anguish of being robbed and killed in his flight, after having murdered his brother, it constitutes a contradiction as it is opportunely noted below when this episode is narrated, or if the episode has only a moral meaning...

In fact, more than answering these questions, which will cease to exist if put in their proper terms, according to theology or faith, in the work of creation, Genesis wanted to state especially:

- the existence of a Creator God of all things, pure Spirit, hence eternal and absolute;

- in the human species, the existence of the Spirit, the Soul, or the Consciousness (the name does not matter but the content of the name) as a gift from God given to one of his creatures, man or the human species, a gift that makes the man, or the species, *in His image and likeness*;

- the immortality of the human species, resulting from its image and resemblance to its Creator, it means, from its participation in the nature beyond time and space, even if limited and conditioned to matter;

- the moral responsibility of the human species, arising from the inherent attribute of Consciousness, from which Freedom necessarily derives, therefore the ability to know, distinguish or practice good or evil;

- the identity, or personality of the woman, for this reason her creation is narrated apart from the creation of man, although based on it, to signify union, love and, simultaneously, the same dignity between him and her, being this episode of separately creation in a way never attributed to any other species.

It is also part of the theological doctrine in this context, the narrative of the misuse of human conscience, which explains the "fall", sin, or disobedience, no matter the name, that is, the deviation of the human species from the natural order established by the Creator through created nature, which gave rise, on one hand, to human suffering (due to the misuse of conscience) and, on the other, to the Project of Redemption, (by the Creator's pre-science of the consequences of the granting of Consciousness as a result of Freedom) essential dimensions of the history of the human species, from its origin and, surely, an object of logic and faith, until its consummation, always realizing that the Bible, whether in the old or in the new Covenant, always revolves around the Redeemer Project.

It is not essential how this happened, because if it were essential, the form, the medium, the characters of the biblical narrative and its consequences, never repeating itself, it would never repeat Freedom, Consciousness and, therefore, all meaning of the human species or of creation itself.

I believe this is also the line of official theology, which is the adequate interpretation of the essential foundations of faith and, therefore, in this perspective, we can walk in the history of the human species, yesterday as today and for the future, in the union and intercomplementarity of science, logic, theology and faith.

b) The creation of Man

In the proposed perspective, Revelation, through the Bible and the theology that interprets it, attribute the creation of man to a special intervention of the Being, whose absolute and eternal spirit we have already analyzed, God, therefore, creator but not created, without limits of any kind, especially of space and time, as Saint John, the already mentioned apostle, also teaches.

The Biblical narrative is well known: "*So God formed man from the clay of the earth, and breathed into his nostrils a breath of life, and man became a living being*" (Genesis 2: 7).

Yet there was in this being an intention in the pre-existing Plan of Creation. The Creator had already decided (I use this expression analogously) to do it, not equal to Himself (which would be impossible, two absolutes), or equal to other living and evolved species, but to do it in His own image and likeness *"let's do the man in Our image and likeness,"* said God. (Genesis 1:26)

The image and resemblance to which the Book refers is evidently not in matter, form, weight, everything conditioned by time and space, from which all things were made. Nor is it in life, already granted to all living beings. In fact, in this Chapter the Bible can be understood as the announcement that God would make a certain species a being, *"a new species made in His image and likeness"*, a purpose which He put into practice as narrated in Chapter 2:7 of Genesis blowing into an individual of this new species, from clay, the figurative form of the matter from which they originated, or were created, albeit by the process of evolution, all species, all living things.

In the case of man, the Bible registers something else, saying that in that image formed from clay, (an evolved being like the others), God inspired a "*breath of life*", and this allows us to say that, in this breath of life, this new creature, coming from the "clay", received something beyond the life of other species. He received a participation that made him *"in the image and likeness"* (Gen.1: 26) of Himself, as He had planned, therefore, independent of space and time, in other words, participant of the spirit, which means, immortal, participant of the eternal and infinite of God.

In this narrative of man's creation is the very foundation of faith, or of the divine or religious meaning of the human species.

d) Considerations on the creation of women.

In the creation of the human species, it must be concluded that it includes the creation of women, not just men. Genesis, however, reports the creation of woman apart from the creation of man (Gen18: 26). This does not occur with any other animal species.

Perhaps the Bible wants to pay special attention to the creation of women, meaning, immediately, that she would not be, in a Creator Plan, just a female, as in the process of evolution of animals, where this creation apart at no time is narrated.

But although, according to the biblical narrative, He made her from man, or through man, surely the Creator wanted to reveal that they are both of the same species, but having an identity of its own, greater than simply attributing to it the functions of a female, as in the other species. Thus, she is a creation willed by God for man, *(It is not good that man should be alone,* Genesis 2:18.) born from him, but created apart, and in this, I believe, it is from the beginning affirmed her autonomy, the woman's own identity and, therefore, her dignity.

The vision of faith, of those who believe in a Redeemer plan, allows us to go further. Perhaps in this gesture, which expresses the dignity of women, the Creator wanted to anticipate what was reserved for women in his Redeemer Plan: the generation of the Redeemer, in the flesh of women, born by His work, in a woman's womb, regardless of man's participation in that generation, because he was born of the Holy Spirit, as we have seen, and we shall better see again in the birth of the Redeemer.

Anyway, other reasons can also be alleged. Perhaps the author of Genesis, reporting the episode of sleep and the removal of Adam's rib to form Eve, gave this episode the strict meaning that keeps another symbol of moral or theological character: the differences between man and woman, but with the same dignity, both creatures of the same species and of the same origin, but different, not equal to cancel each other out, or to repeat each other, but equal in dimension and fulfillment of the diversity of the functions of each, and in their responsibilities.

c) whether the diversity of species includes the human species.

On this last question, it is necessary to initially consider that no experimental science has yet been able to provide an answer. However, this issue has consequences both in the field of social sciences and in philosophy, especially in ethics, law, as well as in theology and faith.

The question is: was there a single initial couple in the origin of the human species, or was there more than one couple, or was there an entire species that evolved?

For those who deny the Creator's intervention in the origin of the human species, it seems normal that an animal species has evolved, in various parts of the globe and at different periods, from its animal condition to a human condition. If so, it would only remain to research in each place and in each layer of the temporal extract of the excavations and, perhaps, we would even arrive in relation to the different species of hominids, perpetuating themselves in their descendants.

In the case of divine intervention, the issue takes on another dimension. I do not know, and I believe that there is not, in the theological interpretation, the hypothesis defended that this divine action has been repeated in several places, creating several couples, at various times, or several places - a hypothesis that, without a doubt, would create new and complex questions.

While the unique creation hypothesis simply has an absolutely important corollary, namely, that this new species, the human species, from the Creative act, would result in its own form and, by a single assumption, of evolution, resulting in the equality of all the members of the species, in all respects, spiritual as well as biological, racial or material, and the DNA differences lately in fashion, do not reach these dimensions. Only marks, or individual differences, not species, are identified.

Multiple evolution, or even the non-existent hypothesis of multiple creation, would give rise to the assumption that there are different human categories within the same human species

This hypothesis would bring with it very serious consequences. Given this evolution by species, at different times, in different places, even of different animal species, it would be inevitable to admit a dangerous argument in favor of difference, or even the superiority of races over other races, as well as in favor of different conditions and identities within of the human species. It can be thought for any of its attributes.

What do the experimental sciences, especially the social sciences, have to say about the diversity of species within the human species and its consequences?

There are questions to be answered by experimental science, or one more Mystery whose answer remains in the field of theology, of Revelation, or of faith. Or the absurd.

2. Good and bad use of human conscience

2.1. The Mystery of Good and Evil

The initial fall of the human species – that is, the evil present since this species has been known – could have been the frustration of the creation project, as a divine work, therefore of the highest good.

In this context, the question arises, the dimension of which experimental sciences such as psychology, sociology and biology, or philosophy, with obvious repercussions on theology and faith, try to answer in vain. This is a fundamental question for all who are aware of the realities of man and his species, or of humanity: the question of good and evil and then of human suffering.

There are several interpretations that seek answers to this Mystery. The current that attributes to evil its own identity, which arises from the duality of the human species, matter-spirit, with matter being the origin of all evil, in a permanent struggle between it, matter, and good, spirit, in a process of continuing purification.

This current has its origin in Manicheism, a philosophical-religious interpretation, preached by Manes, or Manicheus, a Persian priest who lived between the 3rd and 4th centuries AD in Syria, who imagined synthesizing in his doctrine the three great religious revelations: Brahmanism, Buddhism and Christianity. His doctrine spread in the following centuries throughout the ancient world, in Europe especially in the south of France where it inspired some dissidences in the Catholic religion, such as the Albigenses and the Cathars, all condemned by the Church as heretics, in several of its Councils, including Manichaeism itself, solemnly condemned by the IV Lateran Council of 1215.

In fact, by attributing the origin of evil and, therefore, of human suffering to matter, and since matter was created by God, one would conclude that God would also have been the author, or the origin of evil and human suffering.

Manichaeism reached North Africa during Manes' lifetime, towards the end of the third century and had among his disciples a brilliant young man from Ipona who, after converting to Christianity under the influence of Saint Ambrose, Bishop of Milan, came to be become Bishop of Ipona. Died in the year 430, Augustine was declared Saint and Doctor of the Church.

In the exercise of his pastoral munus, Augustine became a severe critic of the doctrine of Manes, expounding in his books and preaching his own interpretation of the origin of good and evil. From St. Augustine's perspective, evil does not exist by itself. Evil is the absence of a good that should exist and does not exist, or is not practiced, or is wrongly practiced. Evil is, therefore, a product of Consciousness, or human choices.

In this reasoning there is no creator of evil, if not human consciousness itself, and its ability to order or practice being - beings and their relationships - being able to order them according to their nature - good, or in disagreement with it, the evil. In this interpretation, God is only responsible for the greatest good that he wanted to transfer, and transferred, to creation: something *in His image and likeness*: the Spirit, the Consciousness that should be used for good.

It is also deduced that, if the Consciousness were taken from man, or if the Creator preferred not to have in his creative work a being *"the image and likeness of Him"*, evil would not exist. But God, even at this risk, preferred to create man, whom he endowed with Consciousness, which means, with the ability to distinguish, or better, to choose, or even better, to do, or finally, to create good or evil.

There is no way not to reflect on the enormity of the dimension, or human dignity, because of the Consciousness and Freedom that arises from it, or how not to reflect, not to understand, or at least glimpse, the dimension of the Redeemer Project, arising from the divine project of doing, through Consciousness and Freedom, a kind of creature *in His image and likeness*, I repeat, the human species.

In fact, it constitutes another unfathomable Mystery, which I would improperly call the stubbornness of the Creator, who reached the limit of simultaneously creating a species endowed with Consciousness and Freedom, and knowing that it could, and would effectively misuse Freedom, He creates another Mystery, the Mystery of a Redeemer Project, Mystery not because of its history, or because of the fact itself, but because of the dimension or depth of the intentions, or the Love of the Creator.

In this Project the whole dimension of His Love lies in having committed a part of Himself, His own Son, who would announce and bring salvation, even at the cost of his bodily death amid human sufferings, although rising to his divine dimension, giving the human species the way and the means to overcome its errors in favor of the good.

Thus, the dimension, value, or price of human Consciousness and its use, and Liberty and its use, or in short, the value, or dignity, of the human species is of immeasurable price in the divine economy.

In fact, it was Pope Francis who said that God gambled on the game of Freedom and, in that game, committed his own Son to the project of Redemption. This Creator's Mystery takes us back to the episode of the mistaken option of the first couple, Adam and Eve, in biblical history. This episode outlines the issue of good and evil in a symbolic narrative in Genesis, according to which, while the first man, Adam, and his wife lived in Paradise, Eve was tempted by the serpent to eat the fruit of the tree of knowledge, which had been forbidden to them by the Creator. She ate it, liked it and offered it to Adam who also ate it and liked it.

They then discovered that they were naked, with the serpent receiving the curse of God and the disobedient couple being expelled from paradise and having them with their descendants to suffer and die, surviving with the sweat of their brows in the cultivation of the land that, beyond its fruits, would produce thorns.

In this symbolic narrative of Genesis, and in the Old Testament, which we will understand better in the final part, it is now possible to understand a little: the episode of sin, guilt, or the original fall and human suffering, narrated by the Bible in its own way, I repeat, wanting to teach that since its origin there was a wrong choice in the use of human Consciousness, or in the exercise of Freedom. Misuse that, with Consciousness, was inherited from generation to generation, and will be inherited as long as Consciousness is inherited, that is, as long as the human species exists: misuse, sin, fall or original stain, whatever the name.

2.2. The Redeemer Project and its origin

On the logical assumption that the Creator loves His creature and has loved it to such an extent that he wanted, through the human species, to have a creature *in His image and likeness*, I repeat, it would not make sense to leave this species without remedy for its errors.

If the Creator knew that it could and would make a mistake receiving this gift, and that the mistake could and would repeat itself, permanently through the generations, He could simply have sacrificed, or taken it away from Consciousness, as punishment, which would probably be the human way to act. That's not what the Creator wanted to do, led by Love to his creature.

He had in his mind from the first moment a Redeemer Plan, which he created not as a consequence of the fall, but in the logic of rescue, of the redemption of his creature to whom "stubbornly", I repeat, an inappropriate term, to give Consciousness and Liberty. This painful, but wonderful Project of Redemption, Mystery conceived from the first moment, if it can be expressed that way, of the Plan of Creation, would accompany the human species, just as the fall or Consciousness would accompany generations, from generation to generation. It is a collective heritage, because Consciousness was given to the species.

This fact of the gift bestowed on the species, however, did not evade individual responsibility for the individual use of Consciousness, or individual choices. In this context of good and evil and the Redeemer Project, it is still necessary to consider a question very close to good and evil, perhaps a Mystery that, just as the fall and the Redeemer Project, accompanies the human species: the existence of good and evil beings.

2.3 - The Angels, or beings of good and evil

There is another Mystery, in the context of the inheritance of good and evil: that evil or suffering entered the world through another species, represented, in the biblical image, by the serpent, the being of evil. We are faced with a new Mystery: that of having existed before the creation of the human species another creature, in the course of the Bible presented as a species in part definitely fallen, which means, evil beings, and another part definitively constituted of good beings.

Such a dichotomy, of a species divided into beings definitely good and beings definitely evil, leads us to conclude that there was a species of creatures that had and lost Freedom.

The Bible often refers to the existence of evil beings, such as fallen spirits presented as demons, or evil angels.

Thus, the Bible, in addition to the original fall to which Eve was induced, the evil spirit appears in dozens of episodes, one of the most audacious being the temptation made to the Redeemer Himself in the desert, as narrated in the gospels and in other narratives in ancient and in the New Testament.

Likewise, the existence of good spirits, angels of God, His messengers, appear in many episodes, from their appearance to Abraham and other patriarchs and prophets of Judaism, to one of the most beautiful episodes narrated in the gospels: the annunciation to the Virgin Mary, of generation and birth of the Savior in her virgin womb, or even, in the meeting of the angel in the empty tomb of the Redeemer, announcing his Resurrection to women.

I believe that this, the existence of a species of pure spirits, good and evil, constitutes another unfathomable Mystery involving perhaps, as has been said, a creation prior to the creation of the human species, a species of purely spiritual beings, endowed of freedom, because otherwise part of this species would not have fallen.

In this case, the price paid by the species was the loss of Freedom, giving rise to necessarily evil beings, those who chose to rebel against God, and necessarily good beings, the messengers or intermediaries of God to human species, protectors of the new species, to whom freedom was maintained despite the fall of the species, although, either because it was conditioned by the limitation of matter, or by the unfathomable purposes of God, among which, the existence of the divine Plan of Redemption of His new creature.

I do not believe, however, how one can go further in understanding this Mystery, even appealing to the Bible, to Revelation, or to theology, leaving only faith.

Finally, among the Mysteries placed in this context of the existence of good and evil, there is a reflection on human suffering and anguish.

2.4. Human suffering and anguish

Psychology, medicine, psychiatry, pedagogy, in short, a whole range of experimental sciences has in vain tried to explain or cure human suffering and anxieties, increasing or standing out, since the last century, among them, psychoanalysis of Freud and the diverse branches arise from his followers.

However, human suffering has accompanied and will accompany the human species, because, according to the biblical narrative, in a way, or in some way, human suffering and anxieties are also explained as a consequence of the wrong initial choice, a consequence of the existence in the human species of the attribute of choosing good or evil. It makes sense, therefore, that, like the initial fall, pain and suffering accompany humanity from generation to generation, pain and suffering and its consequences, and follow the same logic.

The biblical narrative of the initial fall includes its consequences, the expulsion from paradise, suffering and death. Genesis does not take long to show that evil, suffering and death would happen and all the resulting suffering, narrating what happened in the first generation, with two of Adam and Eve's children, Cain and Abel. (Genesis 4:1-8)

Although in Cain's anguish after killing his brother, there comes an extraneous information, if it is referred to history, since, according to the same narrative, by assumption, the world was not yet populated, *Cain reveals the fear of being killed by someone whom he met in the flight to which he was condemned* (Genesis 4:10-14). It remains clear, therefore, in this information its essential objective, that suffering and pain, fear or anguish, must accompany evil.

For this reason, reflecting on this episode, it is seen that evil, misuse, or deviation from Consciousness, constitutes one of the sources of human suffering. Envy, which dulled the murderer's conscience, the same envy, like so many other deviations, pride, presumption, greed, ambition, like Cain's envy, continue today causing pain, suffering and anguish in the world.

In any case, in the project of Redemption, there is also a remedy, or help for overcoming human suffering in faith and in the exercise of its liturgies, especially in the sacraments, and specifically in confession and divine forgiveness of human falls, long before Freud, or the misuse of their consciences, through access to these means.

Thus, from the misuse of human consciences, significant portions of the evil, suffering and anxieties of humanity originate, from those restricted to people or human groups, to extreme conflicts, wars, misery, hunger, persecutions and the deaths, with all the suffering and anguish that accompany them.

Today, avarice, pride, or ambition, that lead to the concentration of wealth, knowledge and its misuse, constitutes one of the biggest causes of the suffering of millions of people, victims of hunger, ideologies of power, racism and even, absurdly, of religious fundamentalisms and of other kinds. However, in addition to the misuse of human conscience, there are other reasons that cause human pain, suffering and anguish.

Suffering can also be a consequence of events in nature, of which the human being is a part, and stems from the limitations of the human species, as well as any nature in constant evolution, therefore subject to the effects of continuous changes that generate consequences.

But even in these cases, although not every natural accident is caused by man, and history is full of facts of this nature, such as the eruption of Vesuvius, the volcano, which

buried Pompeii, today, and increasingly, natural accidents are being caused by the evil and irresponsible action of man on the Planet, the destruction of nature and its resources, the pollution and destruction of the atmosphere, climate change and other events of the same nature.

Man's responsibility for human suffering becomes more serious, especially if we consider that, contrary to the misuse of the development of science and technology to produce evil, scientific and technological advances could be used, in harmony with nature, its preservation and its use for good. However, even in this case, the good use of science and technology, there is no way to avoid the extreme form of suffering, death itself, as an extreme event inherent to matter, part of the human condition.

Finally, under this aspect, one can also attribute to the human condition another kind of suffering more difficult to understand and accept than any other: the suffering of the innocent. Whether the suffering of innocent children, the accused or those who suffer unjustly, the victims of slander, injustice, persecution, etc. To these, the Bible, already in times of the Redemption, has a word from Jesus, in his sermon on the beatitudes (Matthew 5:3-11) *"blessed will you be when men hate you, cast you out, revile you, and when they reject your name as infamous because of the son of man. Rejoice in that day and jubilate, for great is your reward in heaven"*.

Pope Francis, once again, reflecting on the suffering of the innocent, especially children, looking at the Redeemer, Jesus crucified, replied: *all that remains is to look at Him, innocent dead in the midst of unspeakable suffering.*

These are answers to yet another Mystery, that of the suffering of the innocent, for those who believe and so that can also be enlightened those who seek answers. For unbelievers, or for those simply indifferent, there follows conformity with not knowing, or the admission of the absurd. For everyone, each in their own way, the Mystery.

PART 3- THE MYSTERY OF REDEMPTION

1. The Projects of Creation and Creature Redemption

It was seen that, being part of the Creator's Project the creation of a species *in His own image and likeness*, giving it Consciousness and, with Consciousness, Freedom, and knowing, in its pre-science, equally, that, with Freedom that species could choose evil, and would choose, along with this *image and likeness*, He also created a Project of Redemption for this species, so that good prevailed over evil and species, and each individual of that species could use it to walk the path of salvation.

This Project of Redemption, with all the Mysteries that accompany it, constitutes at the same time the reason and the essential content of faith, and in this essential content, with the Mysteries that it brings, this third part brings us to reflection.

1.1. The Project of Redemption in the Old Covenant

According to the Bible, the history of the Hebrew people, chosen by the Creator as the depository of his Redeemer Project, began with Abraham, in the twentieth generation since Adam and Eve, the first couple of the human species, or in the tenth generation from Noah, the survivor of the flood.

Abraham would have lived around the year 1800 BC and was called from Ur, in Chaldea, where he was born, to take possession and dwell in Palestine, where his descendants would constitute a great trusting people of the heritage of the Creator, the one God, and of his promise, the Redemption.

In fact, although Abraham is considered the patriarch, the founder, or the origin of the "people of God", chosen to guard the nature of the true God and His Redeemer project, the Bible divides this time into several periods counted in generations and marked by somehow by God, through direct intervention in its history, through punishments or rewards to the chosen people, as a way to keep his Covenant intact.

The first period would comprise ten generations, from Adam and Eve to Noah, who saved the continuity of the prostituted human species in their customs and in their initial belief in a Creator God whom they disowned, as well as the continuity of the animal species.

This occurred when this just man, Noah, obedient to God's order, built the great Ark where he embarked with his family and a couple of each terrestrial animal species, saving themselves, their families and the animal species. Thus, from the flood that submerged the earth during the 40 days and 40 nights of uninterrupted rain, his descendants would repopulate the earth.

Chapter 9 of Genesis narrates the promise of the Covenant that God would make not only with man, but with all living beings on earth:

- *God said to Noah and his sons: "I will make a Covenant with you and with your posterity, as well as with all living beings who are with you",* (Genesis 9: 8-10), a promise that would be sealed with the Hebrew people 14 generations later, through Abraham, considered the patriarch of this people.

It is important to register, and this is explicit in the following verses, that the covenant extends to all living beings who were with Noah, from which it is deduced that the duty to preserve animals is a biblical commandment for all generations.

In relation to the promise of the Covenant, in the tenth generation since Noah, in the year 1800 BC, God intervened in history, summoning one of Noah's descendants, Abraham, to emigrate from his hometown of Ur to Palestine, where he renewed with God the "Covenant" of fidelity to him and his descendants, which would constitute a great people. The chosen people to keep the one God and the promise of Redemption, as said.

About 500 years later, the descendants of Abraham, chosen to be the great people, also forgotten the Covenant, and ended up becoming slaves of the Egyptians, a slavery that lasted for over 300 years. After this time of slavery, in the year 1170 BC, God, who had not forgotten the Covenant with His people, made a great leader among them arise, Moses, who took the people out of slavery in Egypt and made them return to the promised land of Palestine.

During this return, God used Moses to confirm his Covenant through the prescription of 10 commandments, which should be kept and followed by the chosen people.

In the fourteenth generation, one of Abraham's descendants, David, was chosen by God to organize the Hebrew people into a great Nation, this happening about 900 BC. Until that time the Hebrews lived dispersed in 12 different tribes, and with David they became a great Kingdom which, however, due to the continuous infidelities of the people to the Covenant, had its duration interrupted 500 years later, and the useful classes of the people were taken away as slaves to Babylon, where they would suffer for more than 50 years, while the magnificent temple built by Solomon, son of David, where the Ark of the Covenant was kept, the sacred tabernacle of the ten commandments, was completely destroyed by the invaders, along with Jerusalem itself.

Fifty years after the exile, freed by the will of Cyrus, King of Persia, returning to Jerusalem, they rebuilt the temple, symbol of the Covenant. However, the chosen people still went through a series of conflicts, ending up dominated by the Romans, even before the time of the coming of Jesus, the promised Redeemer, who was not recognized by the ruling elites, with the Hebrew people once again dispersed throughout the world, and its Temple again, and definitively destroyed it, in the year 70 after the birth of Jesus Christ.

It is impossible not to see in this definitive dispersion (not considered the recreation of the State of Israel, without any religious connotation, which occurred in the year 1948 AD) of the Hebrew people, the end of their Covenant with God, replaced by the Redeemer Project that had been consummated with the coming of Jesus.

During this long period that in biblical history would have lasted around 4,000 years from Adam and Eve, or 2,000 from Abraham, the Hebrew people lived between the protection of God and their infidelities, for which they always paid with exile and slavery.

Despite being the chosen people to keep fidelity to the one God and prepare the advent of the Redeemer, despite the continuous preaching of their prophets, who announced that the awaited Redeemer would not be the builder of the great Nation they imagined, Nation that would dominate the world, it – the Hebrew people, or their elites - did not understand that the Redeemer would have a different mission, the Mission of rescuing humanity, bringing it back and, in its fullness, to the Creator's Project for its creation.

However, despite all that the Redeemer would teach and leave as a legacy of a new Covenant, this path of return would not end with his coming, his Message, or his supreme sacrifice. Through the Message, his legacy and his Sacrifice, a New Covenant between the Creator and his Creature would be re-established, thus carrying out the Project of Redemption.

The prophecies of dozens of prophets of the Old Covenant, although in the language of the prophecies, not always transparent to the common understanding, advanced who would be the Redeemer, what His Mission would be, and some of them clearly indicated how it would be carried out. So, they anticipated that the Redeemer would come, what kingdom He would come to bring to every creature and how the performance of this Mission would take place.

If the preaching of the prophets did not help the chosen people or their elites, always dreaming of a Nation that would dominate the world, to understand who the Redeemer would be, how and what He would do, it served to keep them in the hope that a Redeemer, or Messiah would come, which means that, somehow, God would send them a rescuer.

They just did not understand that this release would have an immensely larger dimension than imagined. But the Prophets announced who would be the Redeemer, and there are many Prophets and prophecies. Among the many Prophets, in the text below will be transcribed some prophecies concerning the birth of Jesus and His message and also His death and Resurrection.

To avoid long interpretations, the prophecies cited below, among dozens, are the simplest but most evident, offering enough logic to conclude that the Project of Redemption would not end with a fact, an eventual event, the coming of Redeemer, but it would take place in a process that has accompanied, accompanies and will accompany all humanity throughout its history, from the past, the present and the future, until the end of times.

1.2. In my Father's house there are many dwellings (John 14.2)

By creating the human species whom, through Consciousness, He wanted to make *in His image and likeness*, the Creator knew that the Consciousness, through which He realized the Mystery of the essential participation of His creature, it means, the Alliance between the dimension of matter and spirit, His own dimension *(His image and likeness)* He knew that Consciousness, because of Freedom, could be used, by its creatures, for good and for evil. Surely, the Redeemer Project was born with Project of Creation.

But even when confirming, from the first couple, or since the first generation, the use of Consciousness for evil, the Creator, in choosing the Hebrew people as His chosen people, He certainly did not intend, in the logic of His Love, of letting the rest of his Creatures be condemned to eternal abandonment.

If this logic underlies the redemptive universalization for all humanity through the coming, or the life of the Redeemer, his Message and his Sacrifice, it is surely in the same logic that, wanting to bring forth the Redeemer from the lineage of the chosen people, the Creator has inspired other prophets, who, using their conscience correctly, would inspire other peoples to practice good, even though they did not reach the fullness of the projects of Creation and even less of the Redemption.

This logic leads us to the certainty that the ancient religions, which existed in different ways in all peoples, especially the great organized religions and their prophets, should be understood in this way. Evidently, many cultures or procedures considered as rites or religions cannot be included in this category, due to their deviations, but I remind that some of them, with certainty, inspired consciences for good, among others.

With these examples I want to reaffirm that in the core, in the nature of every human creature, the seal, the mark of a Creator Father was engraved, and this may be an explanation of why, among ancient peoples, even the most primitive, or above all in the most primitives, there has always been some form of religion, or at least some belief in some deity.

From this memory recorded in nature, on the part of many peoples, primitive, sometimes barbaric or cruel forms of divinity remain, or seeking protection from dozens of gods, usually confused with human heroes, or elements of nature, such as the sun, certain trees or animals, the moon or other elements. I don't judge about those.

However, in other peoples, minds, or privileged consciences, purer beliefs were also inspired and practiced, including monotheistic beliefs, often aimed at defining rules of life or behavior, often inscribed in regulatory codes. These characteristics, in any case, do, if not confuse, at least approximate to what can be characterized as religion, or simply as a code of conduct, or even what we could simply call philosophies of life, or commitments of human consciousness, or its subconscious.

Amidst these, among the oldest, and which regulated the life of peoples for many centuries, such as Zoroastrianism in ancient Persia, we can mention others that are still in force, such as Buddhism, Hinduism, or Confucianism. I refer to these feelings, religions, cultures, or religious beliefs, to reaffirm how the religious fact, the relationship of the human species with its Creator, or with a superior Being, or even its Consciousness of good and evil, is present in all times and in all peoples, giving a broader sense to the Redeemer Project, destined, albeit in different ways, to all the people-creatures of God, rather than just the Hebrew people.

Although, according to the Hebrew Bible, or at least the Old Covenant Bible and its interpretations, this Project focuses on the Hebrew people, seeming to want to attribute to them the exclusivity in the Redemption project. In fact, this, the exclusivity, was the greatest illusion of these people themselves.

In the New Covenant, however, the announcement of a universal kingdom, not material, of universal scope of the Message of the Redeemer, was explicit from the beginning in His Message, expressed by the angels when announcing it *to men of goodwill*, (Luke 2:14) without preferences or exclusivity, but in the course of its history, as shown by the apostles of the Message, taking it to all peoples.

In recent centuries, especially, it has given a secure logic to the understanding that the Redeemer Project, in its broadest sense, is aimed at all of humanity, albeit in different ways. This realization, after all, is creating the Universal Kingdom of the Creator's creatures and of their Redeemer, that kingdom that Jesus answered clearly to Pilate when asked if he was a king: *"Yes, I am KING. My kingdom is not from this world. If my kingdom were from this world, My followers would be fighting to keep Me from being handed over to the Jews. But as it is, My kingdom is not from here.* (John 18:33-36).

Remembering that His disciples were Jews, the Redeemer wanted to affirm that his Kingdom was from another world, not another earthly kingdom, but the world of the Spirit, the world of Consciousness, therefore the world of the Creator, a world that in this last century more than in any other, approaches its spiritual and universal identity, with the globalization of civilization itself, assuming, among other values, the universal affirmation of the equality of the entire human species and other universal values.

This statement is confirmed by paying attention to the fact that humanity is beginning to walk in the acceptance of these universal values, further identified as values of a new Mass of Consciousness, compatible with the values of the other realm, brought by the Redeemer to the New Covenant, and also by other ways, or abodes that exist in the Plan of the Creator: *"In my house there are many abodes"* (John 14:2)

I believe it can be said that, in this new phase, the Message moves towards its universalization faster than the Church itself, which remains as the Ark of the New Covenant, guarantee, guard and heir of the essential Message of the Redemption, that does not navigate the waves, but the essence of *the path, truth and life* (John 14:6), serving as a reference and course until the end of time.

Well then, it is this Mass of Consciousness that takes over and grows in the world, source in favor of Peace, Love, Dignity or the equal Rights of all men, Justice, Freedom, Solidarity, Participation, Equality of all, (in the view of faith, all equally children of God) and of the preservation of nature, all values of the Redeemer's Message, or of the essentiality of the Redeemer Project contained in the Bible, in the Old Covenant of the Hebrew people with the Creator and in its evolution in the New Covenant brought by Jesus Christ, the Redeemer.

It seems to me that this is the great step of the Redeemer Project, or the universalization of its Message, as the Redeemer wanted (*"so that all may be one..."* John 17:21)

Having so far sought an approximation of all the facts with the narrative, or the teaching of the Old Covenant of the Hebrew people, I want to focus not only on history, but on the Mystery of Redemption brought by Jesus Christ the Redeemer or the Message by Him brought to all creatures, to all humanity, the Message of the Creator's New Covenant with his creatures, addressed or offered to all peoples by the force of the Redeemer Project.

2. The Redeemer Project and the New Covenant

2.1 Birth and ancestry of the Redeemer, born of God

The evangelist Matthew was among the four evangelists who dealt especially with the origin and infancy of the Redeemer, Jesus Christ, Son of God, second Person of the Divine Trinity. He narrates Jesus' birth in this way: *"Here is how Jesus Christ was born: Mary, his mother, was married to Joseph."* (Matthew 1:18)

The evangelist then says that since Joseph found her pregnancy strange and did not want to be accused of dishonoring her, he planned to leave her. Then an Angel appeared to him (perhaps the same one who had announced to Mary that she, a virgin, would give birth to a son) and said: *"Joseph, son of David, do not be afraid to take Mary as your wife, for what was conceived in her comes from the Holy Spirit. She will give birth to a son whom you will name Jesus, because he will save the people from their sin."* (Matthew 1:20-22)

The realization of the Redeemer Project began to materialize like this, with an incomprehensible story for human understanding, of a virgin who gives birth, contrary to the laws of nature, without having cohabited with any man (and the generation by laboratory would only come more than 2000 years later and, in any case, through the semen and the ovum, respectively from the man and the woman). Joseph, however, who, being a just man, as the evangelist says, with reason from a human point of view, intended to abandon her. But before undertaking his intent, he was convinced, again by an angel, that the conception had taken place by the work of the Holy Spirit.

He believed and took care of Mary and her Son, from His generation in Nazareth and His birth in Bethlehem, His flight to Egypt and during His training, back to Nazareth.

In this way, the prophecies about who would the Redeemer be also began to come true. Seven hundred years before the Prophet Isaiah already stated: *"a Virgin will conceive and give birth to a son who will be called Emmanuel – God in our midst."* (Psalm 7,14)

And about the place of his birth, Micah also prophesied between the years 600 and 700 BC: *"From you, Bethlehem, though small among the tribes of Judah, will come one who will be great among the Nations. Its origins are in the past, in distant times"* (Micah 5,1)

Micah's prophecy about the birthplace of Jesus, in Bethlehem, was only fulfilled because of the census ordered by the Roman Emperor Caesar Augustus, and each one had to present himself in the city of his origin. José and Maria lived in Nazareth, but they were from Bethlehem, from David's house.

On these origins "in distant times", St. Matthew in his Gospel begins the narrative naming the genealogy of Jesus, generation by generation, from the time of Abraham, concluding: *"Therefore, the generations, from Abraham to David, are fourteen, from David to the captivity of Babylon, fourteen generations, and after the captivity of Babylon to Christ, fourteen generations."* (Matthew 1:17)

On the other hand, Genesis equally names the generations, from Abraham to Noah, and from Noah to Adam, so it becomes clear Micah's reason when referring to the origins of Jesus, as *"in the past, in distant times"*. Even considering the symbolic meaning of numbers, as was common among the ancient Hebrews, it is impossible not to see that, since the fall of Adam and Eve, or since the endowment of the creature of Consciousness capable of knowing and distinguishing good from evil, it was in the Plan of Creation the sending of the second Divine Person, Jesus the Redeemer. It necessarily follows from the granting of Consciousness, Soul, or Spirit, that something that resembles the human species to its Creator.

Before moving on to the meaning of the Redeemer's incarnation, I would like to refer to another prophecy of the Prophet Micah concerning the murder of the innocent by the tyrant Herod, frightened by the news given by the Wise Men who, coming from the east and following a star, wanted to find the Child, to adore it.

Fearing that that Child would come to occupy the throne in his place, Herod had all the children born in Bethlehem killed, in the last times that the Wise Men had defined as receiving their message, as the priests had given him knowledge of the prophecy of Micah about the birthplace of the Redeemer, the above-mentioned prophecy in 5.1 which I repeat in a similar version: *"And thou, Bethlehem, land of Judah, art by no means the least among the tribes of Judah, for from thee shall go forth the chief who shall rule over Israel, my people."*

And the prophet Jeremiah, in Chapter 2:18 prophesying the slaughter of children born in Ramah: *"In Ramah a voice was heard, weeping and great lamentations: it is Rachel crying her children; she doesn't want consolation, because they no longer exist".*

To clarify any doubts, historians identify Ramah 3 kilometers away from the entrance to Bethlehem, where the tomb of Rachel, wife of Jacob, son of Abraham and mother of Joseph of Egypt is found.

The fulfillment of these prophecies of Micah and Jeremiah is thus described by the evangelist Matthew 2:3 – 5: *"King Herod was troubled, summoned the chief priests and the scribes of the people, and asked them where the Christ should be born. They said to him: in Bethlehem, in Judea, for thus it was written by the prophet."* They were referring to Micah's prophecy.

Herod then asked the Kings, called Magi, who, finding the boy, would return and let him know who He was, as he also wanted to go and worship Him. The Magi, warned by an Angel of Herod's true intentions, returned by another path. Herod, seeing himself deceived, ordered that all the children of Bethlehem and surroundings, under 2 years of age, were killed.

Matthew writes, showing how Jeremiah's prophecy was fulfilled: *"When Herod then saw that he had been deceived by the Magi, he became very angry and had all the boys from two years old and below slaughtered in Bethlehem, according to the exact time he had asked the Magi. Then fulfilling what was prophesied by Jeremiah* (Matthew 2:16)"

The Child Jesus had fled to Egypt with His father and mother, which allowed, with His return, as Matthew still writes, another prophecy, now from the prophet Hosea, was fulfilled: *"From Egypt, I called My son back"* (Hosea 11:1).

Finally, ending this sequence of prophecies about who would be the Redeemer, with all the circumstances of his childhood, Matthew writes in verse 23 of the same Chapter that Jesus came to live in Nazareth so that what had been said by the prophets would be fulfilled, that *He would be called a Nazarene*, (Matthew 2:23) to signify that he would be despised, as were the Nazarenes.

The story of the Redeemer's incarnation, the second person of the Trinity, taking a human form, which at first glance seems absurd or fanciful, becomes absolutely logical within the Redeemer Plan, from the logic that the sin to which refers explicitly the angel, came into the world through the Consciousness given to the human species by the Creator.

Therefore, logic says that the Creature could only be forgiven, or redeemed effectively (not just apparently) by the Creator. The reason it would be rescued the way it was, is the Mystery on which lean over the prophets, the evangelists, the theologians, in short, those who seek answers...

Certainly, as Redemption was done, the Creator wanted to redeem His creatures, without taking away their Consciousness, or the dimension of Freedom… and in this there is tremendous logic, or the dimension of the greatest responsibility ever given to the human species, its Consciousness and, with it, Freedom.

On the other hand, on the assumption of an incarnation of the Son of God, the second Person of the Trinity, He could not be begotten by a man. But to assume His incarnated divine image He would have to come from the incarnated species (or there would have to be a new creation…) and in this understanding is the whole logic of the Redemption: the Redeemer, with a view to his human dimension, took on the flesh of the organism of a woman like any human being, but with a view to His divine nature, was begotten by God and not by a man, in that woman's womb, therefore, a virgin.

The place of this Virgin in Theology places her as the representative of the human species in the incarnation, a role that was not given to man, because the Redeemer could not be generated by man, being effectively a Child of God. And faced with Mary's astonishment upon receiving the Message that she would conceive and give birth to a son who will be great and will be called Son of the Most High, the Angel explained to her: *"The Holy Spirit will descend upon you and the strength of the Most High will cover you with its shadow. For this reason, the Being that will be born from you will be called the Son of the Most High."* (Luke 1:34-35)

On the assumption that the mission to redeem the human species could only be done as the Creator's work and that Jesus was the Son of God, the second divine person, only by the work of God could Him be generated and the angel, explaining his generation faced with a Virgin barely out of adolescence, involves in that generation a Third Person of the Trinity, the Holy Spirit, the strength, inspiration, the sanctification gift of the Trinity.

All this coming of a Redeemer, among its Mysteries, that of the Redemption itself, the birth of the Redeemer, the consequent existence of the unique nature of a God made man, the realization of so many prophecies, the reaction of the powerful to the birth of a child, it is also a revelation of a one and triune God: The Mystery of the Holy Trinity. The Father and the Son incarnated by the work or action of the Holy Spirit.

On the historical existence of Jesus, it is not necessary to refer to passages from historians or officials in Rome, such as Flavio Josephus, the best known. The certainty of His existence is given to us not only by the arguments that history, or experimental science uses, to prove the veracity of His reports about His existence and His works, or because, despite them and beyond the reports of people with whom He lived, there are His works and His Message, which shaped the civilization of half the world, and which is expanding more and more in universal values. So, I don't see why to doubt the historical truth of His life, His work and His Message.

The same does not happen with other leaders to whom are attributed the origin of religions or cultures, such as those mentioned, Siddhartha Gautama, from Buddhism, Zoroaster or Confucius. These mystics or prophets, in addition to having lived in times where written testimonies did not exist or were barely beginning, only come from other times or from vague traditions that are lost in the mists of time. This is not the case of Jesus Christ. He lived in a familiar age, in identifiable places, in times of advanced civilizations in organization, culture, writing, and other historically accepted elements.

But the essential thing is that His Message, from the beginning, has been changing history, or the world, which occurs most strongly in this twentieth century after the Author was killed and crucified, as I think it has become evident when referring to the end of this book, the growth in the world of Mass Consciousness values, its globalization.

In these new times, the essence of the same Message is closer to a "universal kingdom" opening up each day more to the same messages from "other dwellings" or messages of good that, despite the fall, is marked in the Consciousness of creatures *"of good will"* which, as we have seen, has been announced by the angels in the first news to the shepherds, or to the world, of the birth of the Redeemer. (Luke:2.14)

2.2. The New Covenant or the Redeemer's Essential Message

In the Old Covenant, while the Hebrew people, freed from slavery in Egypt, wandered through the Sinai desert in search of the promised land, which had been occupied by their patriarch Abraham, Moses once again climbed the Mount, where he would receive from God the Tables of the Covenant, containing the Ten Commandments, from which an infinity of prescriptions on worship and on the relationship with God were transferred to the people.

It was about this revelation that the disciples of Jesus, the Redeemer, asked Him what was the greatest of the commandments and had the answer, that to the greatest of the commandments, He also added *all the law and the prophets*, so that there would be no doubt. Certainly, Jesus, in addition to the ten commandments referred to in Chapter 20 of Exodus, referred to all the prescriptions contained in the books of the Bible and even more to the hundreds or thousands of prescriptions of codes and ordinances added by kings or prophets and certainly to the prescriptions of the priests, scribes and Pharisees, those to whom Jesus also refers, who have elephants that blind their own eyes and are left to criticize the specks in the eyes of others.

Surely Jesus' answer can be applied to so many codes, opinions, instructions and other instruments, which were added to His Message and turned into real elephants that blind the eyes to the essentials of what is very clearly, placed in this answer.

Two evangelists narrate the episode of the definition of **the greatest commandment that summarizes all the law and the prophets**: Saint Mark in his Chapter 12: 29 - 31 and Saint Matthew, Chapter 22:36 – 40. Due to their essential importance in the Message of Jesus, I transcribe the two texts.

Writes Saint Mark: *"one of the scribes who had heard Him argue, and seeing that He was answering him well, asked Him: Which is the first of all the commandments? Jesus answered him: this is the first of all the commandments: hear Israel, the Lord our God is one Lord: you shall love the Lord your God with all your heart, with all your soul, with all your spirit, and with all your strength. Here is the second: you will love your neighbor as yourself. Another commandment greater than this does not exist."*

It is impossible not to register the scribe's logic in the face of Jesus' response, noting that this being so, that one should love God above all was fine, but that loving one's neighbor as oneself **would exceed all burnt offerings and sacrifices.** He had then from Jesus, who had realized the wisdom of his observation, the following reply, as if to say: that's right, it exceeds all burnt offerings and sacrifices, but he only observed: *"You are not far from the kingdom of God..."* (Mark 12:34)

St. Matthew, in turn, writes in almost the same terms as the question that the doctor of the law asked him to put Jesus to the test, after a meeting held among the Pharisees: *"Master, what is the greatest commandment of the Law!?"* Jesus answered him: *"Thou shalt love the Lord thy God with all thy heart, and with all thy soul, and with all thy spirit. This is the greatest and the first commandment. And the second is similar to this: you shall love your neighbor as yourself. In these two commandments all the Law and the Prophets are summarized."*

Interestingly, while Mark refers to the fact that Jesus' new teaching exceeded all the Law and the Prophets, Matthew similarly records that the new commandment summarizes everything that was written by the Law and the Prophets.

And speaking of this new teaching, it is impossible not to register what another evangelist, John, writes about the dimension of this teaching in Chapter 13:34-35 of his
Gospel. The evangelist says that Jesus, on the eve of beginning His passion, as one who gives His testament to His disciples, says: *"I give you a new commandment: love one another as I have loved you. In this, if you love one another, all will recognize that you are my disciples"*

The logic of this commandment, which summarizes all the others, including *"the law and the prophets"* must be that whoever loves fulfills all the others, because all the others aim to avoid evil, whether against God, against himself, or against the next. Whoever loves, practices all the commandments.

From so many teachings, so many parables, so many facts narrated by the evangelists, I dwell on just a few passages in the texts with which Jesus marked the Message of the New Covenant, starting with Chapter 5:3-11 of the Gospel of St. Matthew, where Jesus defines, among His disciples, who will be the blessed:

- the meek because they will possess the Earth
- those who have a poor heart, for theirs is the kingdom of heaven
- those who cry because they will be comforted
- those who hunger and thirst for justice because they will be satisfied
- the merciful because they will obtain mercy
- those who have a pure heart because they will see God
- the peaceful because they will be called Children of God
- those who are persecuted for the sake of Justice, because theirs is the kingdom of heaven.

In the end, Jesus left a specific answer to His disciples and to so many others who throughout history, beginning with the nascent Church, would suffer martyrdom and other forms of persecution, because they were his disciples. Said Jesus: *"Blessed will you be when they slander you, when they persecute you, and all falsely say evil against you because of Me. Rejoice and exult, for your reward will be great in heaven, for thus they persecuted the prophets who came before you."*

Evidently such teachings scandalized the powerful Hebrews and opposed everything that were the values of those who exercised power, had wealth, conquered peoples and enslaved the weak, finally those who enjoyed their own power, their own pride, hatred and revenge. For this reason, it was evident that He aroused against Him the envy, hatred and desire for revenge of the powerful.

But this Message was not new. It had been announced at the beginning of everything, already in the song of the angels, bringing to the shepherds the news of the birth of the Redeemer: *"Glory to God in the Highest and peace on earth to men of good will."* (Luke 2:14). I conclude by quoting the first great New Covenant theologian, the Romanized Jew Saul of Tarsus, the greatest convert in early Christianity, St. Paul the Apostle. The Apostle writes in his first Letter to the Corinthians the most beautiful page about charity, not charity as alms, which is explicit in the last paragraph, but charity as total giving to one's neighbor, Love: *"even if I spoke the language of angels and men, if I don't have charity, I'm like a sounding bronze, or a clanging cymbal. Even though I had the gift of prophecy and knew all the mysteries and all the science; even if I had all the faith to* transport mountains, if I don't have charity, I'm nothing. Even if I distributed all my goods to support the poor, and even if I gave my body to be burned, if I didn't have charity, it wouldn't be worth anything."

And after a deeper eulogy about charity, he concludes: *"For now, the three remain: faith, hope and charity. However, the greatest of them is charity."* (1st Corinthians 13:1-13). ³

I believe that the synthesis contained in this set of Redeemer's Messages characterizes the essence of the New Covenant. I stick to them because I believe that Christ effectively taught that Love for God and neighbor, in addition to being the greatest commandment, replaces all the law and the prophets.

But words weren't enough. The Redemption of humanity, where the gift of Consciousness that made the human species *in the image and likeness* of its Creator and therefore, the knowledge necessary to know good and evil and do good, had to be respected, as an essential part of the Project of Creation.

---

³ **Hymn of Charity**

This is the entirety of Chapter XIII of the First Letter from St. Paul to the Corinthians, also called the "Hymn of Charity" according to the direct translation from Hebrew, by a dozen specialists, in Hebrew, Aramaic and Greek, under the auspices of the League of Studies Biblical published in a luxurious edition of Editora Abril, Brazil, 1968. Here is the full text:

"1-If I speak the language of men and angels, but I do not have charity, I am a bronze that sounds or a cymbal that tinkles.

2-And if I have the gift of prophecy and I know all the mysteries and all the science, and if I have all the faith, so as to transport mountains, but I don't have charity, I am nothing.

3-Even if I distributed all my goods, in support of the poor, and gave my body to be burned, if I don't have charity, it doesn't benefit me.
4-Charity is long-suffering; charity is kind; is not envious; charity is not arrogant or puffed up.

5-It does not do what is inconvenient, does not seek its interest, does not get angry, does not take offense into account.

6-It has no pleasure in injustice but rejoices in the truth.

7-It forgives everything, believes everything, waits everything, supports everything.

8- Charity never goes away. Prophecies, on the contrary, will disappear, tongues will cease; science will disappear.

9- Because our knowledge is imperfect and our prophecy imperfect.

10-But when what is perfect comes, what is imperfect will disappear.

11-When I was a child, I felt like a child, I spoke like a child, I thought like a child, but when I became a man, I made what was a child to disappear in me.

12.Now we see in a mirror, in a confused way, but then it will be face to face. Now I know in the imperfect way, but then I will know as I am known.

13-For now, faith, hope and charity remain. But the biggest one is charity."

The Redeemer, not just His own testimony of offering His life in holocaust, left in His legacy the instruments so that the human species, in addition to knowing his Message, had his Consciousness the necessary strength to opt for good, and to do good, being faithful to the New Covenant, and this we will see after reflecting on the supreme witness of the Redeemer: His own life.

2.3. The Redeemer's Supreme Testimony

Sociologists, political scientists, jurists, some will tend to interpret what happened to Jesus after three years of preaching, when He wandered through the villages and cities of Judea and Galilee, healed the sick, raised the dead and planted the New Covenant Message brought to redeem the human species, as a rebellious or revolutionary action.

In turn, at that time, all those who felt attacked, or surpassed in their interpretations, beliefs or privileges, had a reaction of hatred and revenge. Around them their faithful, a small portion of the inflated Hebrew people, accompanied them. Perhaps even some of those who, just a week ago, had triumphantly welcomed Him into Jerusalem. Both did not realize that the Redeemer's Mission had a far greater dimension than a rebel or revolutionary action, or a preaching and an action aimed at replacing the Old Covenant.

They did not understand that the Redeemer had come to perfect the Old Covenant, complete it and fulfill it in the fullness and for the fullness of time.

Like the Hebrews and the Romans who sacrificed Jesus, to think like this is not to realize that this Mission had an infinitely greater dimension and everything that would happen had been predicted by the prophets about the Redeemer, and by He Himself had been announced to His disciples before it happened.

In fact, in His infinite mysteries the Creator, knowing that when giving a limited creature Consciousness and therefore Freedom, was simultaneously knowing the misuse, or its fall that would happen, as this possibility is inherent to the use of freedom.

By absolute logic, equally, it must be admitted that He also knew, and created, from the beginning of Creation, the Project of Redemption and, therefore, the Mystery of the sacrifice, death and resurrection of Jesus Christ, His Son, second person of the indivisible Trinity. This is the ultimate Mystery. The Creator sacrifices himself for His creature... Only the supreme Love... the madness of Love, as interpreted by saints and theologians.

So, I begin by transcribing some Old Testament prophecies and transcribe Jesus' own prophecies about His death and resurrection. I choose among dozens some simpler ones, as they are more direct and do not require lengthy interpretations.

Thus, the book of Exodus in Chapter 12:46, referring to the Passover Lamb that would be sacrificed by the Hebrews, explicitly stated that of the lamb to be sacrificed, no bones could be broken. According to the exegetes, this demand was a prophetic anticipation of what would happen to the crucified Redeemer, when, contrary to what would happen to the evildoers crucified with Him, whose legs were broken so that they would die more quickly, when they approached Jesus, they found that He was already dead and *no bones were broken, only a spear pierced his side, from which blood and water flowed.* (John 19:33-34).

Many are the prophecies about the crucifixion and the torments that the Redeemer would suffer. David's Psalm 21:17 prophesies: *"A gang of evildoers surrounds me: they pierced my hands and my feet."*

And further on, the same Psalm 21:19 continues: *"They divided my garments among themselves and cast lots for my tunic."* Evidently, the soldiers who crucified Jesus and those who divided His garment among themselves and cast lots for His tunic, as it had no sewing, but was woven from top to bottom as the evangelists report, did not know that they were fulfilling prophecies of a thousand years before.

However, the prophet Zechariah in equally ancient times already referred to those who *will look at the One who had been pierced* (John 19:37) and the evangelists confirm those who wept at the feet of the crucified, among them Mary, His mother, the holy women among which Veronica and Mary Magdalene and some disciples among them, of course, John, to whom Jesus gave His mother to take care of her.

Also impressive is the knowledge that the Redeemer had of His next sacrifice, His torments, His death and resurrection, as the evangelists would confirm, each in their own way, transcribing their prophetic words.

Without dwelling on what John writes, from Chapter 16 to Chapter 19 of his Gospel, a true testament of the Redeemer before beginning His passion and death, I transcribe what Matthew and Mark wrote, quoting the prophetic words of Jesus in those days.

Matthew writes in Chapter 26:2 that Jesus, meeting with his disciples, communicated to them: *"You know that in two days' time Easter will be and the Son of Man will be betrayed to be crucified."*

And Mark in Chapter 9:31 transcribing the words of Jesus: *"The Son of Man will be delivered into the hands of men, and they will kill him, but he will rise three days after his death."*
And Mark, again, in Chapter 10:33, also transcribing the words of Jesus: *"Behold, we have gone up to Jerusalem, and the Son of Man will be delivered to the chief*

priests and to the scribes, who will condemn Him to death and deliver Him to the Gentiles. They will taunt Him, spit on Him, lash Him and kill Him. But He will rise on the third day."

These are prophecies of the Redeemer predicting what would happen to Him, for as a Son of God he knew the price He would pay for the Redemption of men, a true Lamb of God to be sacrificed, as Son of men, assumed by the human species, He knew the torments and pains that awaited Him, that brought Him the temptation to resist and made Him suffer in an extreme way, as He would demonstrate in the prayers of the Garden of Gethsemane where, after the Passover supper, he would go to pray: *"Father, if possible, take that cup away from me"* (Matthew 26:39) and even nailed to the cross: *"My God, my God, why have You forsaken me"* (Matthew 27:46).

However, He also knew of the mystery of the Resurrection which He would also go through, while clothed in human and divine nature. As we know, everything that would happen until the Easter celebrations as prophesied and predicted, everything was confirmed by the four evangelists, Jesus' apostles, Luke, Matthew, Mark and John.

With the arrest, martyrdom and death of the Redeemer, and His resurrection, the cycle of the promise of the Old Covenant Redeemer Project ended, so that the New Covenant would have witnesses and be fulfilled, no longer through the blood and death of lambs, or other animals.

In the New Covenant, the witness was the death and resurrection of Jesus, the Redeemer, the "Lamb of God" sacrificed for the redemption of humanity, a witness that would continue through the perennial miracle of Bread and Wine, transubstantiated in his Body and Blood, a miracle instituted by Him, whose supreme Mystery was witnessed by the Evangelists.

In this logic, the bloody sacrifice, witness and beginning of the New Covenant, or of the redeemed human species, would be prolonged with the new bloodless sacrifice of transubstantiation ordered by Him: *"Every time you do this, do it in My memory."* (Luke 22:19).

To leave this testament, of the bread and wine transubstantiated in his body and blood, Jesus wanted His institution to take place on the eve of the bloody sacrifice of His own death, at the Easter Supper of the Old Covenant, with which He would bid farewell to His disciples and that He would become the link between the Old and the New Covenants.

On that day, having fulfilled His Message and knowing that He should return to the Father, as in the last days He had been warning his disciples, John details especially in Chapter 14 of his Gospel, Jesus met them at a supper and before the farewell as, almost repeating the scene and the words, three of the four evangelists narrate: "During the meal He took the bread, blessed it, broke it and gave it to the disciples saying: 'Take it and eat it, this is My body'. He then took the cup, gave thanks and gave them saying: 'Drink from it all because this is My blood, the blood of the New Covenant, shed for many men for the remission of sins'"(Matthew 26:26-28.)

The Evangelist Luke, after repeating the same scene, of bread and wine, adds the final words spoken by Jesus: *"Do this in memory of Me." (Luke 22 :19)*

And Evangelist Mark repeats the same scene and almost the same words. *"Jesus took the bread and, after blessing it, broke it and gave it to them saying: 'Take and eat, this is My body' and then offered them the wine saying, 'this is My blood, the blood of the New Covenant.'"* (Mark 14, 22: 24).

It may seem strange that John, the Evangelist of Love, was silent on this scene.

In fact, the Gospel of St. John was the last of the gospels to be written and John, surely knowing how this Last Supper was repeatedly described, concentrated on five exhaustive chapters, 13 through 17 recording word for word the latest teachings of Jesus, on Love, Union, the painful farewells until they met again with the Father, by whose way He, *the way, the truth and the life* would lead them: *"No one comes to the Father except through Me."* (John 14 :6)

These, the Bread and the Wine, transformed (transubstantiated, transformed substance, says Theology) in the Body and Blood of the Redeemer, would be the new sacrifice, permanent witness in the New Covenant. No longer the sacrificed animals or the bloody sacrifice of the Redeemer made flesh that would take place next, but the same Redeemer present in the continuity of the redemptive work, transubstantiated in Bread and Wine, as before transubstantiated into flesh in Mary's bosom, the extreme Mystery that only makes sense in the logic of Redemption and faith.

Mystery that only makes sense in the logic of Redemption and faith because this simple gesture of farewell at the Easter Supper of a group of 13 people, has not ceased to be repeated, since then, for all centuries and continues to be repeated today, 2 thousand years later, daily in every corner of the Earth. Among those 13 people, present was the Redeemer...

In the logic of the Redemption... and in that logic, it remains absolutely beyond all human comprehension, and for this reason the Mystery is imposed in the integral adherence

to the faith: *Praestet fides suplementum sensuum defectui*, as St. Thomas Aquina says in his "hymn of praise to the Blessed Sacrament"

At the end of the Supper, Jesus withdrew to the garden called the Gethsemani, to prepare Himself and **willingly**, *("Sheath thy sword... do you believe that I cannot call on my Father and He would not immediately send Me more than 12 legions of angels? How then would the Scriptures be fulfilled, according to which it must be so?"* said Jesus to Peter who had unsheathed the sword to defend Him, according to Matthew 26:52-54) **voluntarily**, I repeat, give Himself to sacrifice, fulfilling this way the Redeemer Project, with His supreme sacrifice, His Mission. Arrested, martyred, nailed to the cross, dying, He reaffirmed this fulfillment: *"Father, into your hands I commend my Spirit"* (Luke 23:45) ...and *"all is finished"* (John 19:30).

His Spirit, His divine participation in the Mystery of the Trinity, His identity, had returned to the Father.

His body of flesh *descended to the mansion of the dead* after being taken down from the cross by Nicodemus, who had asked it to Pilate, and having it wrapped in a linen cloth and *deposited in a tomb carved in the rock, where no one had yet been deposited* (Luke 23:53) for the Resurrection as He announced to His disciples, foreshadowing the resurrection of the redeemed human species, on the way back to their Creator.

All of this is for me impressively logic... **and grounds my Faith.**

2.4. The Human Dimension of Redemption, or the New Covenant

Just as the human species could have been saved in other ways by the infinite power of the Creator, the entire human species, now by the sacrifice of the Redeemer, could have been saved instantly, by His Message, and by His Sacrifice (*"...would send me immediately twelve legions..."* Matthew 26:53) how could, from the first fall, have been "saved" having withdrawn the Consciousness, or the Freedom, in the same way as had received it freely from the Creator, who through it wanted to make the species *in its image and likeness*.

But, as we have seen, this was not the Creator's Project. The creature in the *"image and likeness"* was part of the essence of this project and to renounce this similarity, Consciousness, that is, to renounce the ability and responsibility to know and choose between good and evil, it means, freedom, would be to renounce that essence. The way in which the Redemption took to preserve the Consciousness and Freedom of the human species was the life and sacrifice voluntarily assumed by the Redeemer Son and the Message brought to the New Covenant, which was to be preserved and fulfilled from generation to generation.

Thus preserved the essence of Creation that, instead of being saved at the expense of the loss of Consciousness, or Freedom, had its Consciousness preserved and through its options (for good,) the return of the creature, the human species, should take place, to its Creator. In order for this continuity to be full and possible, in addition to the Message and the Sacrifice, the project of Redemption - the Redeemer bequeathed to men the means, the instruments to, knowing good and evil, have the understanding and strength to opt for the good, that is, to fulfill the Creator's Project of building in history, in the place of lost paradise, the Kingdom of Good, of Love and, through this path of Love, truth and life, as defined by the Redeemer Himself, to return to the Creator.

Of these means, or instruments of returning to the Creator, the first was the very content of the Message. The second was the sacrifice of the Redeemer that He wanted to perpetuate with His transubstantiation, no longer in flesh and life, but in Bread and Wine. The transubstantiation (transformation of substance) bequeathed to the world has been the way to remain present among His new chosen people, now in the form of Bread and Wine to nourish the spirit that is in us (the Consciousness, or the Soul).

With this gesture, the Redeemer wanted (or fulfilling the project of Redemption) that the new Sacrifice, the sacrifice of the New Covenant, now bloodless, would be the sacrifice of Love, not of death and would continue to repeat itself, every day as long as human history lasted, anywhere in the world.

This perpetual presence of the Redeemer, no longer in the form of flesh, I repeat, but of Bread and Wine, was explicit, as we can see in Luke 22:19 when He commanded His apostles: *do this in My memory.*

In the logic of the Redeemer Project, of preserving human Consciousness and, therefore, Freedom in His creature, it makes sense that redemption did not take place in a single act, or in a single moment, in the act of the Message or in the sacrifice of the Redeemer, but that happen in the process of people's lives and in the history of the human species. For this reason, in this logic, it also makes sense for the Redeemer to remain among the redeemed creatures, even in the Mystery of the transubstantiation of the Bread and Wine in his Body and Blood, as make sense all the other instruments, said below, left for support and guide this process.

a) The Church, visible sign and Ark of the New Covenant

In the Old Covenant, this support was given through the Ark of the Covenant, the temple in Jerusalem, a whole hierarchy and a whole form of organization of the chosen people, guided by God in its smallest details as shown in the various books such as the books of Exodus, the book of Numbers, the book of Kings and others that also refer to the construction of the Ark of the Covenant, as to the construction of the Temple, as well as to the organization of the Hebrew people and their religious commitments.

In the New Covenant, from a Message preached during three years of the Redeemer's pilgrimage, gathered His Message by the first disciples who followed him, the twelve apostles and, among them, especially the four evangelists, the same tutelage would not be necessary over the new chosen people, the followers of the Redeemer, even because the maintenance of Conscience and, therefore, of Liberty and individual responsibility, was maintained as part of man's association with the Redemption itself.

This line of support and human organization means the institution of other instruments and rites, in addition to the presence of the Redeemer in bread and wine, which would guarantee the permanence of His legacy, as He had promised: *"I will be with you until the end of time."* (Matthew 28: 20).

It also makes sense to maintain a structure, as given to the Hebrew people, its Ark of the Covenant, its Temple in Jerusalem, its hierarchical organizational structure, and as rituals prayer, prayer and animal sacrifice.

In the New Testament, as a structure, the Redeemer bequeathed the Church, in a promise made to Peter, His most impulsive disciple, to whom He promised in response to Peter's affirmation that Jesus was the Son of the Living God: *"You are Peter, and on this rock, I will build my Church, and the gates of hell will not prevail against it. I will give you the keys of heaven..."* (Matthew 16:18)

With the Church constituted by the disciples, around Peter and his successors the entire human organization of the Message, its Ministry and its rituals took place, and in addition to the perpetuity of the Body and Blood of the Redeemer in the form of Bread and Wine, the rituals of the sacraments, recollection and prayer: *"Whatever you ask my Father in My name He will grant you."* (John 16,23)

This logic, the existence of the Church and the Sacraments, took place in the Project of Redemption of the human species, from the moment of its conception. It is in the logic of human things that the Mystery of the Redeemer, God incarnate, made flesh, matter, human consciousness, bequeathed to posterity a sensitive, material organization, a human structure for the support and service of the new Covenant.

Thus, the heritage of the Savior, or the New Covenant, would have two dimensions: the spiritual dimension and the human, or material, dimension. It would exist among men, for men, to serve and guide them, in the midst of their holiness, but also in the midst of their errors or deviations, as an organization, which, even though divine, was given to human hands.

It is significant, however, on these errors or deviations of the Church, to reflect on a significant change in the presence of God in the history of the Church, or His presence in the history of His new chosen people, compared to the history of the Hebrews, of the Old Covenant.

While in the Old Covenant, to whom Redemption had not yet been given, but only the promise of Redemption to be kept, God applied punitive Justice and for each deviation committed, exile, slavery and the time of abandonment came... in the New Covenant, after the Redemption, the punitive justice was replaced by the merciful justice, surely by the merit of the Mystery of the Redemption, transferred to the new chosen people, the followers of its Message. It is logic that the justice of Mercy would prevail over the redeemed human species, as part of the Redemption itself.

b) Some considerations about history.

Like everything else, the structure of the early Church and its hierarchy was very small, but in it emerged the first apostles or interpreters of the doctrine of the Redeemer, the promoters and guides of the communities in conformation, among them St. Paul, the first and certainly the greater to spread the Redeemer's Message, besides among the Hebrews, among all peoples.

Other interpreters of doctrine, on the margins of the popes, who were recognized for the special assistance of the Holy Spirit in matters of doctrine, (*I will give you the keys to heaven...* Matthew, 16-18) have succeeded until today recognized theologians from the time of the older as St. Cyril, St. Augustine, or already in the Middle Ages, Thomas Aquinas and others who founded the doctrine of the Church, as apostolic and universal.

From the earliest days, doubts were resolved in meetings with the Apostles, and the book of the Acts of the Apostles, annexed to the Bible, refers to several of them. Then began to take place the meetings of their successors, who made the hierarchy of the new Church, the bishops, whose primacy was recognized by the Bishop of Rome, successor of Peter, who became the Supreme Head of the Church, or Pontiff Maximus, the Pope.

It was in this city of Rome, capital of the Roman Empire, that Peter, recognized as the first in a line of successors that lasts until today, lived, served as a reference and arbiter of the early Church, and where he gave his life in witness of the faith, among the first of thousands of martyrs, like him, witnesses to the faith at the cost of their lives.

Peter's successors, as happens until today, started to convene new meetings of bishops for debate and decisions on essential questions, the origin of Synods or Councils, although the authority of the Bishop of Rome, the Pope, is recognized even today on Collegiates.

There are no considerable doubts about the legitimacy of this line of successors, but it is admirable that, without the strength to impose itself, or to conquer such a long constant dynastic line of 265 successors, it is already surpassing 2,000 years, uninterruptedly, and it seems, and theologians and historians confirm it, without any deviation in the essentials of the doctrine.

This unity has not been broken in this long history although some differences are little more than rituals like the Orthodox Church, or more distant like Islam, this one with roots closer to Judaism and wandering desert nomads and has suffered frequent internal divisions, some of extremely fundamentalist character. In any case, it is still necessary to mention divisions of greater doctrinal depth from within the Roman Catholic Church itself, such as the Protestant Reformation, also successively divided into hundreds of divergent beliefs, in general, today, called evangelical churches, all followers of Christ, recognized as the Redeemer.

But despite these losses, the longevity and unity that characterize the Catholic Church-Ark of the New Covenant is not common in purely human societies and it alone has more than 1,2 billion followers. It appears that almost 50% of the human species recognizes in God the only Creator, origin of the Universe and of the species itself, and the vast majority also recognizes, in Jesus Christ, the Redeemer.

However, if other sacred manifestations are considered, such as Hinduism, Buddhism, the most representative, this percentage indefinitely exceeds the aforementioned 50%.

It doesn't seem logical to me that more than half of humanity lives in an immense collective mistake.

Later we will also see how the Redeemer's Message in its essentiality is globalized and humanity proceeds in various ways on the path of Redemption, through universal or civilizational values in continuous evolution, not always affiliated with this or that confession.

c) The Sacraments

In addition to the Church, the Redeemer left to support the new Covenant made with the redeemed human species, the sacraments and other instruments of support, of relationship with God, of fidelity to the faith. The sacraments are instruments of Grace (God's presence among men) left to the faithful through the Church, in addition to the supreme Sacrament of the mystery of the transubstantiation of bread and wine in the Body and Blood of the Redeemer, a new way of staying with His chosen people, after having been present in the form of Flesh and Blood, or of the human Person, as we have seen abundantly.

In addition to this Mystery of Bread and Wine, the Church catalogs 6 more sacraments, as specific instruments of the presence, or of divine Grace, in a form and moments essential for the continuity in the life of the community and of each one, in the work of Redemption:

-**Baptism**, by which the faithful enters among the redeemed of the Redeemer, remission of the original fall and adherence to the Message and the Redeemer project.

-**The Confirmation**, by which this adhesion is confirmed before the authority of the Church and infuses in the confirmation the power of the Holy Spirit.

-**The Confession**, by which sins are forgiven, and people are purified to receive in the fullness of peace, the Body and Blood of the Redeemer transubstantiated in the form of Bread and Wine, reconciling with God.

- **The Anointing of the Holy Oils**, intended for the sick, for their cure, their relief from pain, or their passage to eternal life.

- **Marriage**, through which the act of love and the multiplication of the species is sanctified, from generation to generation.

- **The Priestly Order**, or priesthood by which the chosen enter the Ministry of the Church, as ministers of the word and sacraments.

Each of these sacraments constitutes part of the heritage of the Redeemer's Message.

- **Thus, as the supreme Sacrament of the Transubstantiation of the Body and Blood of Christ,** whose circumstance and form of institution was already object of reflection in the Supper in which the Redeemer began His farewell to His apostles, the closest disciples, knowing that in that same night He would begin His sacrifice, each of the other sacraments had its moment of institution.

- **The Sacrament of Baptism**, to which the Redeemer Himself underwent, through John the Baptist who baptized Him in the waters of the Jordan, is narrated in the Gospels of Mark (1: 9 -12) and Luke (3: 21-22) in practically the same terms: *"When all the people were being baptized, Jesus was also baptized. And as he was praying, heaven opened, and the Holy Spirit descended upon Him in bodily form, like a dove; and a voice came from heaven: 'You are my beloved Son; in You I place all my affection'"*.

- The institution of the **Sacrament of Confession** is narrated by St. John in chapter 20 of his Gospel, in one of the most beautiful passages of the apparitions of Jesus to his apostles, after the Resurrection in verses 21 – 23. Here is John's narration: "He told them again: 'peace be with you. As my Father sent me so I send you too.' After these words he breathed upon them saying: 'Receive the Holy Spirit. To those whom you forgive sins will be forgiven; to whom you retain them, they shall be retained'". (John 20:21-23) However, earlier, already in the promise of the institution of the Church and the papacy, the Redeemer had said to Peter: *"I will give you the keys of heaven and whatever you bind on earth will also be bound in heaven and whatever you loose on earth, in heaven will also be loosed"*. (Matthew 16:18)

- **The Sacrament of Confirmation**, in the early Church, complemented baptism through the anointing of the head of the person baptized by the bishop with the sacred oil, as soldiers and fighters anointed their bodies for battles. Although the baptized became part of the Church of Christ through baptism, the sacrament of Confirmation aimed to strengthen the "soldiers of Christ" in faith, in persecutions, strengthening them in fidelity. Although there is no specific mention in the Gospels, this Sacrament linked to Baptism has been a tradition since the early Church.

- **The Sacrament of Priestly Order**, or Priesthood was instituted at the Last Supper when the Redeemer ordered his apostles to repeat in their memory the gesture of transforming bread and wine in their body and blood, the bloodless sacrifice of the New Covenant: *"Do this in my memory."* (Luke 22: 19)

Previously, the Redeemer, as Luke narrates in Chapter 10 of his Gospel, had already induced the Mission that He would leave to His apostles and successors when He sent His disciples to go to all Nations to announce the good news, cure the sick, cast out demons. (Luke 10:1-24)

And still after the Resurrection, in the Gospel of St. John, which describes another moment of the Mission given to the Apostles by the Risen One. I repeat: *"He said to them 'Peace be with you. As the Father has sent Me, so I send you.' After these words he breathed upon them, saying: 'Receive the Holy Spirit, those whom you forgive their sins shall be forgiven; to whom you retain them, they shall be retained.'"* (John 20:21-23).

It is clear in these episodes the Mission that the Redeemer left to His apostles and those who succeeded them, as His representatives in the preaching of his Message, in the ministering of his Grace, through the Ministry of the Sacraments and the word, in the continuity of the Project of Redemption of humanity.

- **The Sacrament of Matrimony**, consecrated since the Old Covenant and, before, since the beginning of the human species, and more, since the creatures multiplied by the union of man and woman, male and female, a union that has always had some sacred meaning in all peoples, surrounded by rituals and prayers.

According to the image used in the Bible, God, having seen that Adam was alone, used all His infinite way of creating, giving him a woman, so that, living together and getting to know each other, they could procreate and populate the earth.

Jesus Christ sanctified the union of man and woman on several occasions, from His first miracle that he wanted to have a wedding commemoration as a backdrop, to answering the question the Pharisees asked Him about the perpetuity of marriage: *"In the beginning he didn't do it God man and woman? For this reason, a man will leave his father and mother and will be united with his wife and they will be one flesh. Therefore, do not separate man from what God has joined"*. (Matthew 19: 4-6) So significant is this expression that a man will leave his father and mother and be united with his wife that it is repeated in the Gospels by Mark (10:7) and again in St. Paul, who dedicated the entire chapter 7 of his letter I to Corinthians to the Sacrament, adding his vision, and also in his letter to the Ephesians (5:21-33), where he describes the entire ideal meaning of the sacrament, placing it in concrete conditions of his environment.

And yet when asked, how then Moses admitted the separation of man and woman, Jesus was harsh in His answer, for He spoke to the Pharisees who were always trying to make Him fall into contradiction. Leaving aside the doctrinal proposition, he went straight: *"Moses allowed it because of the hardness of your hearts"* (Mark 10:5).

However, acting beyond the hardness of hearts, Jesus recognized human frailty, understanding and welcoming at different times women whom the hardness of heart of the Pharisees sentenced as adulteresses or sinners. I mention three of these episodes, where the Redeemer proved that to punishment, he preferred acceptance or mercy as preached by Pope Francis.

- **The first** was in relation to the Samaritan woman who had had five husbands and whom, received by Jesus at the well where he rested, received from Him, in exchange for water, the promise of the water of eternal life and whom, returning converted, ended up bringing all the city to know Him as the Messiah.

- **The second** episode occurred with Mary Magdalene, judged to be a public sinner, who came to where Jesus was at a dinner and, despite the protest of the Pharisees, washed His feet in tears, dried them with her hair and anointed them with her perfume, becoming one of the most faithful followers, accompanying him to the foot of the cross.

- **Finally, the third** episode occurred with the woman caught in adultery and who was brought to Him by the Pharisees to be stoned according to the law of Moses. Jesus, facing them, sentenced: *whomever has not sinned among you, cast the first stone*; leaning down, he began to write in the sand. When He saw that the last of the accusers had withdrawn, He rose and asked: *Woman, has no one condemned you?* And faced with the negative answer He concluded: *No one has condemned you... I do not condemn you either. Go in peace and sin no more.* (John 8: 3-11)

Using mercy, Jesus condemned no one, but converted and gave them peace.

- **The Sacrament of the Anointing of the Sick**, finds its origins in Jesus' compassion for those who were sick, suffering in body and soul, compassion that became concrete in the healing of dozens of sick people, perhaps hundreds or thousands not detailed in the Gospels. Its foundations can also be sought in the order, already referred to, that He gave His apostles to go from city to city to cast out demons, forgive sins and heal the sick, and the apostles, when they returned, happy, told of so many demons cast out, of so many sins forgiven and so many sick healed by the laying on of their hands. But one can also remember the Redeemer's own promise made to the repentant thief that *"today you will be with Me in paradise."* (Luke 23: 43)

d) About prayer and other rites.

Because of what they mean in the process of Redemption, of strengthening the faith and the practice of good, the Redeemer also left reflection, or meditation as a form of union with God, as well as prayer, fasting and also for other good works that He Himself practiced countless times and taught to His disciples, guaranteeing them the validity of different forms of devotion, especially prayer, and prayer in common unity: *"If two of you gather on earth to ask, whatever it is, you will get it from My Father who is in heaven. For where two or more are gathered in My name, I will be among them."* (Matthew 18:19-20)

Finally, as He was praying, a disciple approached Him and asked (Luke 11:1): *Lord, teach us to pray.* Jesus answered him.

"When you pray to my Father, say: Our Father who art in heaven, hallowed be your name, your kingdom come, your will be done on earth as it is in heaven. Give us this day our daily Bread, forgive our trespasses, just as we forgive those who trespass against us and lead us not into temptation, but deliver us from all evil".

After all this, the Message brought, the consummated Sacrifice, the Covenant proposed and the Means or Instruments bequeathed to make the creatures loyal to the New Covenant, the work of the Redeemer was complete, now it is up to the creature putting into practice the Message, repeating the sacrifice of the Bread and Wine, and, using the bequeathed means and instruments, to return to the Creator.

This return takes place, and can be reflected, from the individual point of view, which begins with birth and is completed in the death of each person, or it can take place and be considered in the history of the human species, that is, in the construction of a world – a medium, a culture or a civilization, according to the same Message and the same Redeemer Project, brought to the salvation of creation's favorite species: the human species.

3. The Creature's Return to the Creator

3.1. The fundamentals

I transcribe one of the last teachings of the theologian and priest Paulo Bratti, who expressed himself in this way about the path of the human species returning to its Creator, to whom He gave the mission of taking care of all creation, dominating it, preserving it and using it for his sake.

In this theologian's teaching, the Creator's respect for the Consciousness and Freedom given to His creature to fulfill this mission of returning to the Creator is clearly stated once more, not only in his individual acts, but in the construction of a world or a medium where men live, it means, from a culture or a civilization that is the construction of their own way back, or back from the creature to the Creator. Says the theologian: *"If man is not autonomous and responsible for his actions, he is no longer a man. He can and must therefore (being responsible for his actions) create his own human milieu, that is, of civilization and culture as its economic, political, cultural and moral language (or structure). If God intervened at the level of human initiative, He would compete with our Freedom".*

If, on the one hand, man, by his individual actions, walks towards the return to the Creator or in the opposite direction, if, on the other hand, he builds a world (**his environment, his culture or civilization**) according to the plan of God, it is to build the path for this return, so from this perspective, there is no way not to initially reflect on the different angles, or paths of this return, within two perspectives:

- **about one's own histo**ry, the history of each one, that is, Redemption as an individual process;

- **about the participation of the human species in the Old and New Covenants, today in the construction of globalized Civilization**, analyzing the history of the Redemption or, in a broader sense, the participation of the entire human species in this history, 20 centuries after the coming of the Redeemer, His Message and His supreme sacrifice.

3.2. Redemption as an individual process

The initial question, from the individual point of view, is whether the creature is born with a tainted Consciousness from a fall in the origin of the human species, or whether this inheritance constitutes an individual taint linked to the gift of Consciousness and Freedom, being attributed to the individual only because he belongs to that species that misused them. The reality is that, according to tradition and theology, it is through baptism that, regardless of its origin, or its nature, this stain is erased.

It also makes sense that, since this stain is erased by baptism, the first of the sacraments to be received as traditionally practiced and according to the teaching of theology, it makes sense, as said, that baptism should be administered soon after the birth of the child, freeing it of any inheritance of evil, leaving the Chrism or its confirmation, as a religious option to be ministered later, after the so-called age of

reason, because from this age on, children add conditions of free, individual option. In any case, it can still be registered that John the Baptist has baptized adults, including Jesus Christ, although, evidently, the circumstances were different in relation to the post-Redemption era.

Finally, in this understanding, it also makes sense that the sacrament Confirmation has this name, because Baptism is confirmed under the blessings of the Holy Spirit, in terms of the ritual, as it was in the baptism of Jesus, the Redeemer, and assumed by the interpretation of theology and the liturgy of faith.

Within the meaning of Redemption, that it takes place in history and not just in the act or in the life of the Redeemer, it is understood that individual redemption does not take place in a single act either, be it baptism, or any other eventual sacrament, or just in obedience to one or another commandment, but it is carried out in the continuous practice of faith and its concrete implications, from the moment the option for faith has been made.

Thus, in history, as the Redemption takes place since the participation of the Redeemer and will continue to happen until the end of time, in the same way, it has to be the individual redemption, of the people, from their option for faith, it is better if this option comes from the beginning, although fidelity to the act of option matters more in a process that lasts until the individual end than, properly, from the moment when the option began. What matters above all is fidelity in the faith, which can be cited as an example in Paul of Tarsus, converted as St. Paul the Apostle, who testifies: *"I have fought the good fight, I have finished the race, I have kept the faith. [8] Now there is in store for me the crown of righteousness, which the Lord, the righteous Judge, will award to me"* (Letter to Timothy 4:7-8). He embraced the faith in adult life, even after having, as a Roman centurion, strongly persecuted Christians.

In fact, the idea that it would be a greater respect for freedom to leave the option for faith to adulthood would require from a rational point of view, as a prerequisite for an adequate choice, a minimally in-depth study of religions, the option for a certain religion, or the denial of any religious option. However, in practice, this study, or preparation, as a rule does not occur, and tends to decrease more and more from generation to generation, because this loss has been aggravated by this same cause, from generation to generation.

Finally, it is worth considering, from the point of view of faith, that, since Baptism, as it is, the first of the sacraments, this delay deprives generations of the Grace granted by it and by the other sacraments, leaving them without access to it, Grace, for all that time, in ignorance of the very reason for being of the Sacrament, that is, of the support and strengthening of faith, or of Grace, as well as of the necessary strength for strengthening at the time of the choices regarding the good or to evil, which is the meaning or *raison d'être* of the sacraments and of faith itself.

At last, I believe, and competent pedagogy confirms it, that the introduction to the religious issue, as to other important issues, could not fail to be part of the educational process from early childhood, because among these important issues, faith is a fundamental element in definition, especially of values, but not only. It is also a basic instrument in reference to ethical and moral issues inherent to all human actions, an instrument that will accompany people throughout their lives. It would make no sense that these essential choices should be taken without the option, or at least without the basic knowledge, about matters of faith and its values.

This is a dimension referring to the history of faith, individually considered. It so happens, however, that culture, or civilization, although it has its own identity, are also both the result and consequence of what the people who build them are and are the result of the identity of what these people are, what they think, what they do. Therefore, it is not adequate to build society, whether it be called culture or civilization, without the participation of the options defined by the people.

3.3. On the participation of the human species, in the Old Covenant, in the New Covenant and in the construction of Civilization

From all that has been said, from what can be better understood and from what constitutes and continues to be the Mystery embedded as an essential reality in all essential questions, it is surely confirmed that the process of Redemption as a social fact, it is deduced that, necessarily, this process has two other dimensions:

- **A divine dimension and a human dimension.**

There is no doubt throughout the analysis that the Creator wanted the Project of Redemption to be carried out, respecting these two dimensions. For this reason, the Creator of all things, at the moment of creation, by the breath of Consciousness in His creature, wanted it to be, as we have seen, in its image and likeness, therefore, that it participates in an attribute of its nature.

In sequence, the Redeemer would not allow the redeemed creature and its organization, or form of relationship, to cease to be an essential part of the same process, left only as a divine responsibility. The human species was created to associate with the Creator in the continuity of the creative work, which did not end with the creation of the species or its redemption, but which will continue until the end of time, or the arrival at the Omega point, yet referred to, characterizing the New Covenant of the Creator with His creature.

This leads us to reflect on how the Redeemer wanted to make the Mystery of the creature's participation in this process happen.

a) The participation of creatures in the Old Covenant

Using the biblical view, in the Old Covenant, the Book shows us how the participation of the chosen people took place. To them was given by the Creator the promise that the Redeemer would be sent, making this people witness or guardian of the existence of one God, Creator of all things. This was the essential of the form of the creature's participation in the Redeemer process, in the Old Covenant.

At the same time, however, the Bible shows us how much and how many times the Hebrew people, chosen, failed in fidelity to this Covenant, in function of what, in the same way, it shows us how much and how many times the people were punished for their infidelities. This series of punishments, which begins with the expulsion from Paradise of the first couple, with the anguished flight of Cain and continues with the Flood, even before the formalization of the Covenant, which would only happen with Abraham, this punishment of the flood, in which the human species, decimated along with all species of land animals, was saved from extinction, because there was a just and faithful man, Noah. Thus, by Noah's faithfulness, the human species was saved, and successive generations were saved.

However, as narrated in the same Bible, later on God has not found a minimum of righteous people, through whose faithfulness the perverted cities of Sodom and Gomorrah would have been saved, cities that, for their infidelity, were destroyed by God's punitive justice, and on them He caused fire to fall from heaven, according to the biblical narrative.

Also, for the same infidelity, the people, rescued from slavery in Egypt, for 40 years were left to wander through the Sinai desert, so that the entire generation born in Egypt and saved from slavery would die, including Moses, their leader, without entering the Promised Land.

Finally, for the same infidelity, the majestic temple in Jerusalem built by King Solomon was destroyed, and the Hebrew people, especially its elite, were led to fulfill 50 years of slavery in Babylon, from where they were freed by the emperor Cyrus, impressed by the preaching of prophet Daniel, remembering and prophesying the waiting time and the coming of the Redeemer.

However, even after the coming of the Redeemer, the majestic Temple was once again destroyed, rebuilt by the Hebrews after their return from Babylon. The new destruction by the Roman armies would take place in AD 70, after the people screamed in Pilate's courtyard, condemning the Redeemer to death: *"His blood is on us and our children."* (Matthew 27:25)

With the destruction of the Temple and the city of Jerusalem itself, the Hebrew people, chosen by the Old Covenant, was spread throughout the world, so that they would no longer be recognized as the depositary, or the sole depositary of this Covenant. The Redeemer had come to seal the New Covenant, replacing the one for whose promise the Hebrew people, by their faithfulness, and despite their infidelities, had been chosen as guardian. Such was the process of participation of the Hebrews in the Project of Redemption, according to the Old Covenant.

How would the participation of the people be, and which would the people be, in the project of Redemption, or in the New Covenant?

b) **Participation in the New Alliance.**

In the New Covenant brought by the Redeemer, the commitment would no longer be with a people, or a Temple, but with the entire human species, for what would be carried out by the Sacrifice of the Redeemer, His Message and His inheritance, which would unite all peoples until the end of time.

With the Redemption, therefore, the other face of the image of the Creator God was revealed, from the punitive justice of the old Covenant, which gave way to the God of merciful justice, the God of Love as was explicit, and this has already been analyzed previously, in the Message, in the Sacrifice and in the inheritance left by the Redeemer, including in the assistance that, according to the promise to the apostle Peter, would be present in his Church until the end of time (Matthew 16: 18).

This new Covenant, however, would not eliminate by itself, through the sacrifice of the Redeemer and the other gifts of the Redemption, all the errors, the falls, or any infidelity, which would continue to happen to a redeemed humanity, in that sequence.

In the New Covenant, as a sign and instrument of the Redemption, the Redeemer was immolated for everyone, His Message was left, and also, as instruments to strengthen and preserve fidelity in the faith, the sacraments, prayer and other signs were left, as we have seen.

For all this heritage left to humanity by the sacrifice of the Redeemer, the Mercy of God, in the New Covenant, was extended to all humanity, despite its infidelities, whether infidelities of the individual faithful, of peoples, of the "Ark of the New Covenant", or of the human species. Mercy, therefore, becomes part of the essence of the New Covenant, which is evident, but what meaning would the coming of the Redeemer, His Message, and especially His Death have?

In this context, the human dimension of the new guardian of the faith, or Ark of the New Covenant, the Church, given to human hands, its errors and infidelities, happened in its long history and the original doctrine remained permanently preserved. It is included in this context, according to theology based on the Pope's infallibility, under the special assistance of the Holy Spirit in matters of doctrine, fulfilled the promise, already referred to above, made to Peter by the Redeemer. This assistance is logical, on the assumption that, for the process of Redemption, the Redeemer was committed to death.

I don't know if the logic of the sciences itself can explain that, despite the errors and the constant or eventual, private or collective breakdown of the Alliance, or of the very depository of the Message and its instruments, the Church as the Ark of the New Alliance, the human species, two thousand years later, continues walking along the path of truth and life, the path left by the Redeemer, or by the theology of Love, personified in the Projects of Creation and Redemption.

On this path, with stumbling and slow, but continuous steps, the human species walks... and opens paths for its individual and collective return to the Creator. It so happens that the existence and saving force of the Mystery of Redemption makes the difference between science and relationships in the Old and New Covenants, instituted by God with the human species, now redeemed from its infidelities by the sacrifice of the second divine Person made flesh, who assumed and wanted to carry on His shoulders the sins, or errors, of mankind...

Through this Mystery, essence of the Project of Redemption, the merciful God of the New Covenant must be better understood, and it is also better to understand that so many evils occur in the world, without the corresponding punishment fallen from heaven as, according to the biblical narratives, happened in the Old Covenant and still so many expect it to happen in New Covenant times. But they won't.

It took two thousand years on this journey, with fidelity and infidelity, without the tiring of God's Mercy towards the new chosen people, the whole human species, strength and mercy never so necessary and to be sought at this moment in the Message of Redemption, when the humanity, Civilization or Culture referred to by the theologian Paulo Bratti, mentioned above, grows, becomes increasingly complex, globalizes and becomes more and more interdependent, or threatening, due to the conditioning of force and power that to this globalized society they impose, or condition, the advances of Science and Technology.

c) **The Redeemer project in globalized civilization**

For all this, the advances of Science and Technology reflected in this analysis of the Creator's relationships with His creatures, induce us to deduce, and reality proves, that the options of human Consciousness, and therefore of Freedom, in the face of these advances in Science and Technology not only remain, but assume a new and infinitely greater responsibility, forcing us to ask the question:

Is the human species ready to take on this new dimension?

It is always necessary to reaffirm, and for this reason I reaffirm, that, in His Mysteries, the Creator, in the project of Redemption, preferred to let Consciousness

prevail in the human species and, with Consciousness, Freedom, putting the human species again in every moment before the option between good and evil... This tension between good and evil as well, and in a special way happens today and will happen in the future, especially due to these advances in Science and Technology, with the human species being responsible for the consequences of how one might use it.

It is, therefore, in this new, complex and constantly changing society, yet redeemed by the Message, the sacrifice of the Redeemer and its heritage, that each one, individually, or human institutions collectively, is constantly faced with these same options, between the good and evil. In this continuous option, **the whole logic of this process that underlies my faith**, gives me confidence in affirming that, despite the errors and deviations, the values and the Message of Redemption grow and will continue to grow in this civilization in continuous change.

This Message grows, in the first place, by discovering, approaching, and seeking the unity of the Redeemer's Message, with equal or similar values and Messages coming from other times, from other sources, or from *other dwellings of the same Father*, as it was seen previously.

I think it makes sense in the richness of this perspective, considering the Catholic Church, the uninterrupted succession of the 266 popes since the Apostle Peter, the first Pope, and the Christian people, as the guardian of the inheritance of the patrimony of the Redeemer or, as has been said, essentially as the true Ark of the New Covenant, guarding, guaranteeing and referring to the Redeemer's Message, in ecumenical communion with other Messages that inherited and live the same values and were certainly also the object of the Redeemer's Message and death.

It is in this context, of a society that is globalizing, that I want a reflection, albeit minimal, on the last six popes whose mandate I had the opportunity to know and live in greater depth, realizing how, conductors of the "Ark of the New Covenant", they, in the midst of resistance and misunderstanding, insert it into the course of this globalized civilization. For me, this vision was, and it is, of special importance, in order to admit, or accept, the special assistance of the Holy Spirit, or the Divine Trinity, in the construction of the path of the human species, or of every creature, in search of abodes of Father.

In my view of this process, started by Pope John XXIII, son of farmers in the Veneto, the one typified in the name of Cardinal Roncalli, nothing intellectual, but who, with the unexpected convocation of the Second Vatican Council, surprised the world, opening the Church to society in transformation, *"aggiornando"* the Ark of the New Covenant, in the certainty that they would count on the assistance of the Holy Spirit in this work... and it couldn't even be different. Without a doubt, this Council was, in this effort for *"aggiornamento"*, a significant milestone to continue harmoniously inserting the Redeemer's Message with the transformations of civilization so that it, the Message, can continue to be light to illuminate the paths of the return of the human species, finally, from every creature to the Creator.

John XXIII's successor, Paul VI, a first-rate intellectual, codified the Conclave's decisions, being succeeded by John Paul I who, in turn, only bequeathed a fraternal smile to the world and in his shyness was taken to the bosom of God, just 33 days after taking office...

At that time, the church needed a hurricane similar to Paul the Apostle who in the early Church traveled the world and preached the Message to Greeks and Romans, and to make this similarity effective, the Holy Spirit brought Cardinal Wojtyla, John Paul II, from distant Poland, who carried the Council to the world, preached to the Powerful and the Poor, subjected regimes and renewed the Church as the Conclave wanted, or perhaps God,

the Holy Spirit. John Paul II was replaced by Benedict XVI, again an intellectual, but who, lost in the midst of transformations and feeling unable to respond to the complexity of the transformed world, as well as the flawed structures of the Vatican bureaucracy, resigned, opening a new space for the Message, as the Holy Spirit wanted.

Also coming from the end of the world or, as I said, coming from among the people, where the Holy Spirit went to look for him to the surprise of the world, more than to "upgrade" the doctrine, which had been done by the Council, the Cardinal Bergoglio, who chose to be simply Pope Francis, came to put into practice Love, poverty, acceptance, mercy and understanding for all men, as the Redeemer did, whether saints or sinners, not putting himself as Judge of all people, of all religions, of all races, but all called by the Redeemer's Message and redeemed by the Mystery of Redemption.

In this understanding of the essentiality of the Redeemer's Message in a globalized world, pluralistic in the plurality of human consciences, their Freedom and the resulting responsibility, Francis, the Pope, seeks the construction of justice and peace, only possible in a social organization, with a new economy (said by Francis, from Argentina or Assis?), open to the participation of all human beings, as Jesus, the Redeemer taught and practiced, in respect for nature, as the human species is also a work of creation, and thus has practiced and nailed in his letters, among which it is essential to register the Encyclicals, *Lumen Fidei*, *Fratelli Tutti* and *Laudato Si*, (The Light of Faith, We Are All Brothers, and Praised be You... Lord) the Messages of Faith, Charity or Love and Praise to God (throughout all nature) summarizing the essentiality of the Redeemer's Message, a response to the anguish and hope of the world, this transformed world.

The succession, characteristics and work of each of these Popes clearly reveal to me that the Church, Ark of the New Covenant, receives at every moment the divine presence, carrying out, in accordance with the promise of the Redeemer, that He would be with His Church until the end of time, (Matthew, 28-20) sustaining her in fidelity to the doctrine and despite, or because of, her human frailties, her falls or her mistakes in things, which also existed, and history bears witness.

I know that some do not like this interpretation, but it has always been like that, as from the beginning some did not accept the Redeemer, and his redemptive Message, but it is there, illuminating the world.

So, the Church returns to being the Ark of the New Covenant, guarantor of permanent values and new universal values, *in My house there are many dwellings* (John 14: 2), Love as the greatest commandment, Peace, human rights, respect for nature, accepted and lived in this fraternal communion, which makes all one, as the Redeemer wanted, (*so that all may be one*, John 17: 21) and in this way the Message of salvation extends to every creature, all peoples and all nations, despite the enormous dimensions of the path still to be traversed.

In fact, it is necessary to distinguish the essence of the Message and how it is lived. But it is necessary to realize that, as a result of this transformation of the world in this Global Village, many barriers that separated religions, ideologies, people, groups, cultures, races, peoples and nations, businesses and identities, were, are and will continue to be broken down by the evolution of Civilization itself.

As a result, it becomes inevitable to respond in an adequate, complex and harmonious way, on how to build and walk together with the human species, on this path that remains to be traveled, *the path of truth and life*, that is, the path of Love, *in which everything is summarized, law and the prophets,* as the Redeemer affirmed and wished that the Omega should be the place of arrival, as well as the Alpha of departure.

3.4. Amorized Creation

It was a surprise to me that the path taken in making this book, attentive to experimental science, but progressing in the application of the logic of philosophy that took me to the horizon where theology, or Revelation and faith, made me reach the same horizon or place of arrival, to the same Omega where the path I took and the place I arrived in my book For a Participative and Solidary Civilization. THE PROPOSAL: **the construction of a Civilization or a loving Creation.**

However, in that book I was led by the studies of anthropology, history, sociology, and economics, after more than 50 years of studies, experiences in public life and academic life, which led me to the conclusion that the process of evolution of humanity walks towards the construction of a Participatory and Solidarity Civilization, which I called AMORIZED CIVILIZATION, an expression that, by the way, I looked for in Teilhard de Chardin, cited in that as well as in this book.

For this surprise, I repeat that by these two different paths I arrived in the same way to the Faith which, in its essence, could simply be called the construction of LOVE, the Love that sums up *all the law and the prophets* (Matthew 22:36-40) a Love that, surpassing the context of Civilization, inserts the human species, humanity of all times, or all of Creation, in the Redeemer Project, or in the vision of faith, which takes us back to LOVE, the Creator God of all things. I am sure that only by reaching this transcendental dimension, absolutely essential, the Amorized Civilization becomes possible, and it is in this absolute sense of the concept that it can be built.

Anyway, it is still for this same surprise that I want to dwell on the identities of the conclusions of this book, The Mystery of Everything, especially with the book For a Participative and Solidary Civilization, THE PROPOSAL, (2019) but also with the previous ones, among them, PARTICIPATION AND SOLIDARITY, The Third Millennium Revolution (2004) and THE AGE OF MAN, Foundations for a humanized civilization (1982).

These books identified, in the current stage of Civilization, a great number of values that coincide with the values of the Message of the Redeemer, anxieties and pursuits, these values, which I have given the name of Mass of Consciousness.

The Mass of Consciousness constitutes an identity that unites people and groups or institutions in increasing numbers around the world, or throughout the human species, in these times transformed by its characteristics of communication, relationships and interdependence, which transforms everyone into inhabitants of a new village, which is new because of its global dimension, absolutely different from medieval villages or from other times.

Those who live in this new Village and perceive it, or assume it in its complexity, members of the Mass of Consciousness, are receiving, spreading and living, consciously or not, the essentials of the Message: peace and harmony among people, groups and Nations, justice or human rights, which make all human beings equal (*all children of the same Father*, adds faith), solidarity and cooperation, respect for nature and the preservation of the Planet's natural resources, or of the

Universe, finally Love, which surpasses all law and prophets according to the Message, the same Love, which is called to inspire all the structures and all the actions of the new Civilization, according to the essential aspirations of the human species: the Mass of Consciousness.

This identity, regardless of any characteristics, which could separate rather than unite people, groups, in short, Civilization, is actually calling the human species, or the humanity of these new times, to live in harmony in diversity (because it is this complexity that characterizes the evolved nature) and, in the view of faith, in the construction of paths to return to the Father, regardless of ideologies, beliefs, races, customs, regimes or any other attributes.

In short: the values of the growing Mass of Consciousness in the world, those books summed it up in the expression Participation and Solidarity, whose essence, in turn, is identified or synthesized in LOVE. Love, which, according to the essential Message of the Redeemer, summarizes all the law and the prophets, and which, consequently, constitutes the essence of faith and its commitment in the construction of all amorized Creation.

It is in this harmony between the Message and the Mass of Consciousness that lies the security of building a new organization of the human species after the advances in Science and Technology, that is, part of Creation: **The Amorized Civilization.**

Returning to the analysis of what justifies my Faith, I must conclude that, if **LOVE** was the Creator principle, which gave rise to the human species and that, having been lost from the beginning by the misuse of Consciousness and recovered through the Redeemer Project, it is clear, in this perspective, that the end, the arrival, or simply the objective, the target of the arrow, to use once again the language of Teilhard de Chardin, the factor that moves and makes the history of men, or creatures, survive, it must always be **LOVE**.

If this were not so, the project of Creation, whose misuse of Consciousness was known to the Creator by virtue of his temporal knowledge, if it did not include the Project of Redemption, it would actually have been a suicidal project, as it would be Love allowing that in Love was inserted the germ of its own and ultimate destruction.

The Project of Redemption, therefore, justifies Love not only as the origin, but as the end, the goal of all **Creation and Amorized Creation**, or building the Amorized Civilization, therefore, means to give continuity and fullness to the project of Creation, in permanent and continuous evolution.

3.5. That all them may be one (John 17:21)

But... are there signs that the human species is heading towards amorization?

There are so many signs of this path that it is difficult to start by citing this or that one as the greatest sign, or the most important. But the perception of the existence of the growing power of the Mass of Consciousness is, without a doubt, the first factor that must be noticed, which does not always happen, because the values that make it up in essence, or sometimes because they are just unconscious, escape more easily to quantitative or material perception... waves that are more easily perceived than the depth of the ocean...

When, however, the memory arises that only a few centuries ago slavery was an institutionalized mark, as was the society of castes of all kinds, which denied or ignored any condition of dignity as an essential right of all who belong to the human species, or perhaps to every creature, and that in this century this dignity, expressed essentially in human rights, has already become a global conscience, when this memory is raised, as said, one can see the dimension of the step reached: human rights as a natural right and a universal command that extends to all human beings, or to every creature.

I know that in practice, neither in their scope nor in their universality, human rights, or the dignity of every creature, are respected and promoted. But this resistance does not take away its dimension, on the contrary, it reinforces the unanimous demands and the universal clamor of the Mass of Consciousness and the path still to be taken, in the awareness that the fight for good will happen while the human species exists, that this is the meaning and dimension of the Message itself: *I am with you always, to the very end of the age.* (Matthew 28:20)

Another fact, that history, sociology, or the simple observation of things that happen every day, visible even in the media, although not always explicit in their meaning, show us the advances of the Mass of Consciousness.

In the still recent past, people, the masses went to the streets to demand revenge, against people and against any other groups, or to demand war against other nations, other peoples, other regimes, other countries and, it is true, there are still remnants of these attitudes in various parts of the world.

However, today, the same media shows us that, in other parts or in almost all of the world, and with increasing frequency, people and masses take to the streets to reject everything that affects human dignity and to demand coexistence between peoples, between races, regimes and ideologies, in short, to demand peace…

When the current, newly elected Pope Francis visited Brazil, crowds were seen, the media and research institutes estimated it to be in the order of more than 3 million people, gathered on Copacabana beach to watch religious rituals or listen his word as the supreme guide of the Catholic Church, even though it is considered that an incalculable number of those people neither belonged to the same faith, nor even belonged to any faith at all. What was seen was the Mass of Consciousness present…searching for the words of hope, the support for the manifestations of faith and its values, the message of Faith.

Due to the extreme importance of the Mass of Consciousness in the construction of the new world, or of the loving Civilization, I will dwell a little longer on this reflection of such essentiality in relation to the Message of Redemption.

a) Mass of Consciousness and its meaning.

- **The occurrence of crowds** during the visit of Pope Francis can be considered natural in a country like Brazil, considered the country with the largest Catholic population in the world. But when you consider that on his visit to the Philippines, from a minority Catholic population, millions of people also gathered for the same visit, coming from 32 countries, from the most diverse religions, from Buddhism, Islam, Hinduism, or religion none…

- **When you think, in contrast**, of how many millions of people were killed because of the religion they professed, and residually, how many still die because of remnants that survive such conflicts, despite the international rejection and enlightened segments of their own beliefs, ideologies or cultures, which condemn such remnants of the past that insist on repeating themselves in the present…

- **When you think** about how many religious leaders, institutions or communities formally seek in a global movement of ecumenism to live in harmony and mutual respect, in the diversity of religious beliefs and forms, in a significant movement in favor of peace, solidarity between ideologies, races, individual options or between communities, peoples, origins, etc., religion, or the most diverse forms of beliefs turned into a bond of union and solidarity, evidencing the richness of diversity and union...

- **And yet, when one can think** that in this context of search and coexistence- *so that everyone is one*, it is not allowed to fail to register messages, which can be considered true prayers in the anguish of non-encounter, such as *"make love, don't make war"*. As this prayer arose, circumstances show that even in its desperate path, a desperate generation demonstrates to the world the eagerness to love of all peoples, of all generations, in this complex and confused society, changing, transformed and transforming...

- **Finally, when one realizes** that nations have moved institutionally to get closer to each other and bring their peoples closer together, resolve any disputes or interests peacefully and live together in the multiplicity of forms of participation, solidarity and peace instead of war...

- **When you think...**

...**you just have to open your eyes** and see the way you've been, realizing that there's still a lot to be done, to be able to keep walking...with certainty, more than just hope, that doing what remains to be done is possible and, because of this certainty, take another step and so... step by step...

b) On the way to Unity

However, in addition to walking step by step an individual path in the search for Love, or to assume a responsibility to contribute to the construction of a new civilization, amorized, participative and solidary, it is possible to assume that this is not just a dream, a demand, a moral or a religious vision of human processes. It is, in fact, about building the survival of the human species itself and the environment in which it inhabits, especially aware that a stage of evolution has been reached in which the human species has developed instruments and forms of behavior that can lead to its destruction or, on the contrary, to unimagined levels of human life and coexistence.

If these threats exist, and they do exist, it is all the more necessary to realize that, in addition to the growing strength of the Mass of Consciousness, there are initiatives that work towards cooperation, solidarity, harmonious coexistence, in short, Peace and Love, opening the eyes to its transcendental meaning, a dimension usually not perceived even by the initiatives and their initiators.

Initially, I would like to mention two institutions with a global dimension, together with thousands of public and community institutions, which are dedicated to spreading the construction of unity in diversity, or in the direction of loving Civilization, as you wish to interpret.

I believe that this perception implies giving these Institutions, in addition to a political or social dimension, an ethical, human, or civilizing dimension, the dimension of the Creator's Plans for His creatures... It is necessary to perceive, and affirm, that **a society, or a Civilization in accordance with the values of the Redeemer's Message is part of the very project of Creation and its evolution... you must understand! I reaffirm...**

In order for these dimensions to be perceived, I first cite **the UN-United Nations Organization** which, with the difficulties inherent in complex and innovative social processes, brings together the 193 national States in a global forum for debate on world problems, for dialogue and articulation between the Nations, promoting and supporting actions of international cooperation, mediating conflicts, defending Human Rights, acting through bodies that influence the political, economic, cultural, financial, production and business areas, in favor of peace, greater equity and justice among Nations, in addition to other actions aimed at what I called unity in diversity. I know that, on the one hand, its goals are more advanced than the realities of its members, which relativizes them... but it is a step and, step by step...

So, I think because for me it is evident, and I believe it is absolutely true, that the objectives of the UN are values in tune with the values of the redemptive Message or with the Creator's Plan for His creatures, having, therefore, a dimension much greater than a simple institution of political order or material interests, national, international or of any other kind.

The great obstacle of the UN, in this perception and in this context, is that it exists and has to act in a society whose members are organized, live and still act according to the principles that organized the Civilization of the first so-called industrial revolution, or capitalism financial, or on the other side, of the socialist proposal. Meanwhile, Science and Technology changed the world. Institutions need a step of equal size, one more step...forward.

However, there are no signs that the UN can free itself from the interests that condition it, positioning itself in the direction of the transformations brought about by the advances of Science and Technology and the values of the Mass of Consciousness in the organization and in the relationship between Nations or, even more, in the proposed perspective, of a participative and solidary society or, even more, of an Amorized Civilization... and, **returning to the vision of my Faith**, on the path of Return of the Creature to the Creator.

This would be more than a step. It would be a journey, a civilizing turn. In any case, I consider it absolutely valid and perhaps necessary to affirm the UN as a link in overcoming so many barbarities and in the direction of the new society, the new Culture or the new Civilization, which I call amorized, a call from the human species, especially if considered the values that it, the UN, represents.

I refer, then, to the **European Union, the EU,** an institution that must be seen and understood in the same dimension and meaning, for being the pioneer of an exemplary model of social, political and economic integration, integrating 27 Nations of the European Continent.

These same nations, in the first half of the last century, were involved in two world wars that resulted in more than 100 million deaths, not to mention the incalculable rates of physical and social destruction caused to their own countries and to countries around the world. These painful events highlighted the tragic consequences of wrong choices, the result of competition and conflicts in relations between peoples or between States and Nations. But surely, they also served to awaken consciences to the importance of a harmonious and cooperative relationship between peoples and social organizations, or a new conscience among people, generating a new form of human coexistence, also towards overcoming barbarism and construction of a loving civilization or, in reference to my faith, towards the return of the human species to its Creator.

Exceptions, however, continue to exist in the world, suffering localized conflicts, such as those arising from the rearrangements that followed the fall of the Soviet Union, as well as other residual conflicts from the old or current colonialisms or even caused by localized disputes of a character ideological, racial, cultural or religious, but no other conflict had proportions even comparable to the aforementioned conflicts, which took place before the formation of **the EU and the UN.**

In the case of the European Union, it is also important to notice significant support programs for the poorest countries, as well as other common decisions on issues that would have previously led to new conflicts, while the inhabitants of each country, who now have supranational citizenship, had greater freedom of movement, cultural exchanges and other encounters, in a continuous evolution of human coexistence, breaking down borders and walls that separated them and in favor of the same universal Consciousness, of liberation and freedom in the course, I repeat once again, of a humanized, cooperative, participative, in short, amorized civilization.

However, support and cooperation programs are still mainly due to African countries and other depressed regions of the planet, often a consequence of the domination or exploitation of countries that have accumulated and continue to accumulate the wealth of the world, not just countries of the European Union, but of the new colonizers of capitalism, whether state or corporate, who produce and maintain regimes of concentration and exclusion that are strengthened by the monopoly of advances in Science and Technology.

It is necessary to register, with the necessary vehemence, that this new form of exploitation is the same that occurred in the past due to territorial occupation and predation of resources in exploited territories. Today, that same exploration continues to take place through the systems of concentration that dominate the world, whose origins and methods are questions that have not yet been put to the judgment of history, including the sustainability of Civilization. But, if the necessary transformation does not occur in a timely manner, this denunciation will happen, there is no doubt.

In addition to these advances in the social order towards transformation, it is also necessary to mention the universal awareness that grows, beyond the cultures or the type of civilization in which we still live, in favor of the Earth's ecology or preservation, especially of its threatened natural resources, at the moment when humanity developed instruments capable of destroying the Planet, its waters, its climate, its fertility, its atmosphere and threatens to lead to pollution and destruction to space, as has already been warned in another moment...

It is necessary to bring to the conscience of the world a deep reflection, courage and truth, on the preservation of the Planet, and on the sustainability of this civilization that destroys it.

This is a cry from the evolving human species - the Creator's project, evolution, to which political, economic, religious, or any other order must respond.

This cry is to become stronger every day, as the question is placed in the dimension of faith, and in this perspective, it is necessary to go beyond, and think about the survival of the universe and everything in it, admirable creature of God, the Creator, given to man to take care of everything and to complete his work.

However, I want to warn about the threat of rolling back from the steps taken.

Regarding the examples mentioned, especially about the European Union, it is necessary to be aware of proposals, movements or objectives, which intend to return to the old nationalist quarrels, including a return to competition between interests of all kinds, including among remnants of old privileges that still survive. This also means the threat of simply seeking the transformation of the Institution into a bloc whose essential attention is turned to competition with other countries or with other blocs.

If that happened, the European Union and other institutions that followed the same path would no longer be the step forward they represent, towards a participative and solidary organization, or towards a loving civilization.

As when I referred to the UN, I want to say that the EU and its model of organization and relationship also have a much larger dimension than the small cyclical dimension or occasional interests, to assume a civilizing dimension, in the dimension of building a new Civilization, as I said, amorized, or **in the vision of faith**, I repeat, the return of the human species to the Creator.

But for the globalized human Civilization to assume this dimension, it is necessary to strengthen the Mass of Consciousness, making it aware of the urgency of recreating the entire political, economic and social organization, transforming the current competitive, concentrating and excluding civilization into content, essence and form, to make it participatory and solidary, that is, amortized, I repeat. Globally amortized, I repeat and complement.

This is the civilization defined by the book I have already mentioned, Participation and Solidarity, and especially by its synthesis, THE PROPOSAL, which, as I said earlier, to my surprise, through other sources and other methods, in other words came to this same conclusion: the need and urgency to build, in love, the new Civilization, the Civilization of the post-technological era.

From the perspective of faith, which is the perspective of this book, only this transformation will complement the search for full coherence between the organization of the human species and its functioning, with Civilization redeemed by the Redeemer, amidst the transformations brought about by the advances of Science and Technology.

In addition, I want to note that, as they were mentioned, it is not only the European Union or the UN and their bodies that, consciously or unconsciously, are part of these projects, of Creation and Redemption.

Other initiatives, institutions or manifestations are positioned along this same path: the opening of the Catholic Church and other Churches to live in ecumenical union; the repudiation of war, racism and other forms of disrespect for human rights; multilateralism in relations between Nations and harmonious coexistence in the diversity of people or cultures; pluralism and the disinterested help of institutions such as *Médecins Sans Frontières,* or efforts to reduce excessive inequalities, so that distributive justice is effectively achieved, in short, the defense and cultivation of freedom. These are all signs that the essence of the Redeemer's Message, directly or indirectly, consciously or unconsciously, constitutes an essential part of the construction of a new format, or a new organization and way of coexistence for the human species, or the new stage of Civilization.

It is this set of institutions and manifestations that indicate and give us, as I have said, more than hope, the certainty that, despite the setbacks, we are living, in history, a fertile moment of the divine project of Creation and the project Redeemer of the return of the human species to Love, Alpha and Omega of all that wonderful Mystery in which it fell to us, human species, to be Consciousness.

3.6. The creature's return to the Creator and the ultimate destiny of the Universe

Given this fertile moment in the process of the human species in evolution on its way to an Amorous Civilization, one more step in the Creator's Project for His creatures, a moment in which the immense universe is being penetrated, up to the mathematically imaginable limits, and of matter, penetrated to the last chains that constitute it, at the end of these **reflections on the reasons for my faith**, considering these reasons, within my limitations regarding the experimental sciences, complemented by the logic of philosophy and illuminated by the interpretations of theology, and facing, finally, the supreme Mystery of Revelation, I return to the initial question and its part in the Mysteries of Everything: **where does the human person and its species go, and where does the universe and all Creation go?**

What we see, what we can analyze, in the history of evolution, regardless of the acceptance or not of faith, everything allows us to say that the human species and civilization walk towards overcoming dependence on purely material realities, seeking the path of prevalence of the spiritualization of matter, or the prevalence of spirit,

or of Consciousness and its attributes, over matter and its use. Consequently, I allow myself to say that the entire Creature, together with the human species, advances along the path of its own evolution and the evolution of its organization, that is, of people, of their Consciousness and of their forms of relationships, in the direction of Love.

Finally, that the Creator, who wanted to make the human species in *His image and likeness*, also wanted that, through this species, the entire universe delivered to it (or would
have delivered only a part of the Universe, Planet Earth, or the System...) was taken back to the Alpha, the Love of the beginning of everything, through the Omega that identifies with it, the Love of the arrival.

This view is an essential part of the reasons for my faith and, in my understanding, makes it perfectly in harmony with experimental Science, at least the one within my reach, with the logic of philosophy, the interpretations of theology and Revelation.

This return to the prevalence of Consciousness, or Spirit, in the human species, despite the Mystery that is found in everything when, beyond the surface, essences are sought, becomes more understandable, considering that the species was created already possessing in itself that *"image and likeness"* of his Creator.

Yes, the human species and its essential Mysteries, but what about the Universe?

I conclude with a final reflection on these two Mysteries, that of the Human Species and that of the Universe, meaning, therefore, all Creation.

- **admitting** that the human species was created when the Creator blew, into matter properly prepared or evolved, something *in His image and likeness*, something similar to Himself, to His attributes, among which Spirit or Consciousness...

- **admitting** that the gift of Consciousness constitutes the very gift of the Spirit to the human species and, therefore, Consciousness is, by nature, essentially different from matter, as it has no shape, weight, mass and, consequently, there is no way it can die, or, in other words, considering that Consciousness is by nature immortal...

- **it becomes imperative** to conclude that immortality in the human species imposes itself from the moment it was created with something *in the likeness and image of the Creator*, call it Spirit, Soul, or Consciousness. It shows an enormous truth, therefore, the office of the dead when it says that life is not taken, but transformed: *Vita mutatur, non tolitur.*

This means that death, in fact, is the liberation of the Consciousness, or of the Soul, or *of the image and likeness*, despite its essential unity with matter. This can be the trauma of death, separation from the Spirit, great Mystery, in fact, the smallest part of the great Mystery, which reaches its real dimension when it connects with the integral resurrection of the human species, which imposes the resurrection, in some way, of the matter itself, so that the integrality of the person is preserved.

If the Spirit - the Soul, or the Consciousness - is immortal, there is nothing to resurrect in it, for it does not die. It breaks free. But its completeness must be resurrected in it.

Logic, however, allows us a reference to the death and resurrection of the Redeemer, who exclaimed in His agony *"Father, into your hands I commit my spirit"* (Luke 23-46) as His body descended to be buried like the body of any creature, although in His case He was buried, in the perspective of His divine Nature and the Redeemer Plan, to be resurrected on the third day.

The image, I believe fully applicable, in the first part - the Spirit, the Soul, or the Consciousness of creatures liberated from matter, returning to the bosom of their Creator, the sky, or not returning and not returning would mean the frustration of existence, the Hell!

But what about matter, abandoned by the Spirit? The Redeemer's body was also resurrected on the third day and there is the testimony of all the gospels about his Resurrection, and his *"ascension to heaven"*, which means, in his case, to have returned to the Being
of his divinity, therefore to his full essential identity.

However, in principle, it seems that the human soul, the Consciousness or the Spirit, separated from matter, from the body, would not be complete if it lost its identity as a human person. Unless you find your fullness in God, despite your incompleteness.

But how will the resurrection of the body, destroyed by the abandonment of life and the Spirit, take place? How will the body be raised? Surely the resurrection of bodies will not be matter, I mean, at least the same matter...

There remains one more final Mystery to be answered by theology, or, in addition to human answers, kept among the Mysteries of the Creator, revealed or not, and, among men, **the object of faith. Of Absolute Faith.**

We have already seen the case of Lazarus or the son of the widow of Nain or the daughter of Jairus. But they were resurrected to continue living their life in this world, which is not the case with the resurrection of the human species.

In relation to the Universe, the Mystery also remains, referring to the duration or end of matter, mass, form, time and space, from which the universe is made.

It seems that nothing prevents matter from being maintained indefinitely by the power of the Creator, I use the word indefinitely instead of eternally and, keeping proportions, scientists will conclude, or imagine, as they have proposed, forms of perpetuity or succession of universes, expansion, concentration and restart. However, the Mystery remains and is beyond man, beyond faith-referring to the duration of matter and, for the time being, remains in the domain of speculation of experimental science.

No. We don't know everything. But we can safely say that the universe will not be an immense machine that spins without meaning, an immense clock mechanism that no longer keeps time, or that is immobilized in its dimension, lifeless and motionless. I do not see, therefore, that there can be answers to everything from experimental Science, Philosophy and Theology, leaving only **Revelation or Faith**.

However, it is necessary to state that, being the human species the only one to whom the Creator, having given Consciousness when He wanted to make it *in His image and likeness* and having loved it so much that he created a Redeemer Project to redeem it in its downfalls and strengthen it in the use of His image and likeness and, finally, having made for it the immensity of the Universe, **it is necessary to state that there is no way not to get down on one's knees in respect of the greatness, dignity and responsibility of the Human Species, and of each Person, individually,** since it is up to itself, to each person, to give meaning and soul to all Creation. On the other hand, it also shows the enormity of **sin, or crime, of any form of disrespect that is committed against the Human Species or against any Person.**

But if there are other species of conscious beings in this immense Universe, and it is possible that there are, then so much more, at the stage and in the place we are, the Universe would be, for the human species, an absolutely impenetrable Mystery, because the human species would be reduced to its own little world and perhaps absurdly involved in the wars between the worlds, which has only been, until now judged as fiction, as so many other things were judged before they happened... But how would it be, in this case, the whole story of Creation, of its Consciousness - the human species, presupposition of freedom and, consequently, of its Redemption?

Everything would have to be rethought, from zero point, even if astrophysics reached beyond the limits of this small world, fine chemistry unraveled the last secrets of matter and genetic biology surpassed the mysteries beyond the strands of DNA and RDA bringing new questions of immense dimension for human life and coexistence, morality, ethics, law or, finally, expanding the human Mysteries, **the Mystery of Everything**.

More than ever, **in this immense Mystery, the human species can only have Faith.**

However, whatever the advance in the search, beyond this immense Mystery, **the essential Mystery** must still remain, and that is why, equally, in this essential beyond, only limited, **essential Faith** must have meaning and place. that we went through matter as the Creator wanted to make us.

The Creator, about Whom the wisdom of St. Thomas Aquinas referred, affirming that, in relation to God, we can better understand Him for what *He is not than for what He is.*

What He is, we will know in our return to Him. For now, because of the Mysteries, Faith remains, and Faith gives us the assurance that we will one day reach the fullness of Revelation.

Osvaldo Della Giustina

From my retreat at Villa Aurora, inner Brazil, in 2020.

INDEX

PRESENTATION 9
THE MYSTERY OF EVERYTHING
(From Experimental Sciences and Philosophy to Revelation and Faith) 13
INTRODUCTION 13
PART 1 – THE MYSTERY OF THE UNIVERSE 21
  1. Of the infinity or eternity of Universe 21
    1.1. What Experimental Sciences say 21
    1.2. Will there be an answer to the new questions? 24
    1.3. Where are we going? 35
    1.4. Where does the universe go? 37
  2) The Answers of Theology, or Faith 41
    2.1. The assumptions of the answers 41
    2.2. The origin of the universe 44
    2.3. About the nature of the Creator Being 46
    2.4. About the origin of life 49
PART 2 - THE MYSTERY OF THE HUMAN SPECIES 55
  1. On the Origin of the Human Species 55
    1.1. What do the experimental sciences and logic say? 55
    1.2. The Narrative of the Bible and the Interpretation of Theology 58
  2. Good and bad use of human conscience 64
    2.1. The Mystery of Good and Evil 64
    2.2. The Redeemer Project and its origin 67
    2.3 - The Angels, or beings of good and evil 68
    2.4. Human suffering and anguish 69
PART 3- THE MYSTERY OF REDEMPTION 72
  1. The Projects of Creation and Creature Redemption 73
    1.1. The Project of Redemption in the Old Covenant 73

   1.2. In my Father's house there are many dwellings (John 14.2) 77

2. The Redeemer Project and the New Covenant 80

   2.1 Birth and ancestry of the Redeemer, born of God 80

   2.2. The New Covenant or the Redeemer's Essential Message 85

   2.3. The Redeemer's Supreme Testimony 89

   2.4. The Human Dimension of Redemption, or the New Covenant 95

3. The Creature's Return to the Creator 105

   3.1. The fundamentals 105

   3.2. Redemption as an individual process 106

   3.3. On the participation of the human species, in the Old Covenant, in the New Covenant and in the construction of Civilization 108

   3.4. Amorized Creation 115

   3.5. That all them may be one (John 17:21) 118

   3.6. The creature's return to the Creator and the ultimate destiny of the Universe 126

KNOWLEDGE OF EXPERIMENTAL SCIENCES OR PHILOSOPHY DOES NOT CONSTITUTE FAITH.

FAITH IS THE ACCEPTANCE OF THE KNOWLEDGE REVEALED, OR THE MYSTERY.

FAITH IS THE OPTION TO BELIEVE, TO ACCEPT, EVEN THOUGH IT IS NOT UNDERSTOOD, OR EXACTLY BECAUSE IT IS NOT UNDERSTOOD.

(from the introduction to this book)

Osvaldo Della Giustina

Graduated in philosophy with a course in Journalism and a postgraduate degree in Strategic Planning and Human Resources.

Secondary and higher-level teacher of philosophy, sociology, education and history.
He implemented and directed institutions like: Fundação Educacional, now Universidade do Sul de Santa Catarina (UNISUL); University of Tocantins Foundation (UNITINS); Santa Catarina Work Association (FUCAT) and Santa Catarina Association of Educational Foundations (ACAFE).

As a public man, he was Secretary of State and State Deputy in Sta. Catarina, Secretary for Higher Education Reform and Chief Secretary of the Office of the Governor of the State of Tocantins, in Brazil.

In the federal area, among other functions, he has been Assistant to the Press Secretariat of the Presidency of Brazil and Chief of Staff of the Ministry of Education, the Federal Council of Education and the Ministry of the Environment and the Legal Amazon.

Born in Orleans, Santa Catarina, Southern Brazil, on July 21, 1936.

(EAR 2)

OSVALDO DELLA GIUSTINA's Biography

Author of 17 published books, including books on literature, sociology, economics, philosophy and strategic planning. He has books published in Portuguese, French and English, which can be found in graphic edition or on virtual platforms. Between them:

THE AGE OF MAN, Foundations for a New Social Order (1982); REFLECTIONS ON EDUCATION (1989); Mystical Biography of Sta. Betinha: THE GIRL OF THE ANGELS (1992); Humanization of Society, THE REVOLUTION OF THE THIRD MILLENNIUM-I (2000); PARTICIPATION AND SOLIDARITY - The Third Millennium Revolution II (2004); INSTITUTIONS ALSO HAVE SOUL, the Unisul History (2012); THE NEW UNIVERSITY in a Changing World (2017); For a Participatory and Solidary Civilization - THE PROPOSAL (2019). He has also published hundreds of articles and lectures in Brazil and abroad.

www.ingramcontent.com/pod-product-compliance
Lightning Source LLC
Chambersburg PA
CBHW051103230426
43667CB00013B/2418